T0192663

Electronic Conduction

Classical and Quantum Theory
to Nanoelectronic Devices

Textbook Series in Physical Sciences

This textbook series offers pedagogical resources for the physical sciences. It publishes high-quality, high-impact texts to improve understanding of fundamental and cutting edge topics, as well as to facilitate instruction. The authors are encouraged to incorporate numerous problems and worked examples, as well as making available solutions manuals for undergraduate and graduate level course adoptions. The format makes these texts useful as professional self-study and refresher guides as well. Subject areas covered in this series include condensed matter physics, quantum sciences, atomic, molecular, and plasma physics, energy science, nanoscience, spectroscopy, mathematical physics, geophysics, environmental physics, and so on, in terms of both theory and experiment.

Understanding Nanomaterials
Malkiat S. Johal

Concise Optics: Concepts, Examples, and Problems
Ajawad I. Haija, M. Z. Numan, W. Larry Freeman

A Mathematica Primer for Physicists
Jim Napolitano

Understanding Nanomaterials, Second Edition
Malkiat S. Johal, Lewis E. Johnson

Physics for Technology, Second Edition: With Applications in Industrial Control Electronics
Daniel H. Nichols

Time-Resolved Spectroscopy: An Experimental Perspective
Thomas Weinacht, Brett J. Pearson

No-Frills Physics: A Concise Study Guide for Algebra-Based Physics
Matthew D. McCluskey

Quantum Principles and Particles, Second Edition
Walter Wilcox

Electronic Conduction: Classical and Quantum Theory to Nanoelectronic Devices
John P. Xanthakis

For more information about this series, please visit:
www.crcpress.com/Textbook-Series-in-Physical-Sciences/book-series/TPHYSCI

Electronic Conduction

Classical and Quantum Theory to Nanoelectronic Devices

John P. Xanthakis

CRC Press
Taylor & Francis Group
Boca Raton London New York

CRC Press is an imprint of the
Taylor & Francis Group, an **informa** business

First edition published 2021
by CRC Press
6000 Broken Sound Parkway NW, Suite 300, Boca Raton, FL 33487-2742

and by CRC Press

2 Park Square, Milton Park, Abingdon, Oxon, OX14 4RN

© 2021 Taylor & Francis Group, LLC

CRC Press is an imprint of Taylor & Francis Group, LLC

Library of Congress Cataloging-in-Publication Data

Names: Xanthakis, John P., author.
Title: Electronic conduction : classical and quantum theory to
   nanoelectronic devices / John P. Xanthakis.
Description: First edition. | Boca Raton : CRC Press, 2021. | Series:
   Textbook series in physical sciences | Includes bibliographical
   references and index.
Identifiers: LCCN 2020028630 | ISBN 9781138583863 (hardback) | ISBN
   9780429506444 (ebook)
Subjects: LCSH: Electric conductivity. | Nanoelectronics. | Quantum theory.
Classification: LCC QC610.4 .X36 2021 | DDC 537.6/2--dc23
LC record available at https://lccn.loc.gov/2020028630

ISBN: 9781138583863 (hbk)
ISBN: 9780429506444 (ebk)

Typeset in Minion Pro
by KnowledgeWorks Global Ltd.

To Milta and Liana

Contents

Part III Devices

Preface

Our world is filled with electronic gadgets such as the radio, the TV, the mobile (or cellular) phone, the laptop computer, not to mention the complicated sensors and actuators of the heavy industries. At the heart of all these appliances lies only a very limited number of types of electronic devices: mainly the resistor, the capacitor, the diode, and the transistor. Every electrical or electronic engineering course is not complete without either an initial or a deeper understanding of these devices and there are indeed excellent books on these. However, in the last 20 years we have entered the nanoscale region of electronic manufacturing in which the workings of these devices are no longer dictated by classical transport theory but by quantum transport, i.e. the electrons must be treated as waves as opposed to particles. A significant amount of knowledge of quantum mechanics and solid state physics is needed to comprehend quantum transport. There are excellent research monographs on the subject of quantum transport but most of them assume a fair amount of knowledge, which students may not have. On the other hand, present electronic devices textbooks barely touch quantum transport. This book attempts to bridge this gap. It is intended to be self-contained so that the student has no requirement of an extra book on physics or materials science, although students of these subjects aiming at applications can also profit from this book. The task is not easy because a thorough understanding of modern electronic devices requires a knowledge of the energy bands in semiconductors and this, in turn, requires an initial knowledge of quantum mechanics. All these are included in this book. However, many devices still work by classical conduction and the educated engineer needs both pictures. After all, one must be convinced of the inadequacy of classical transport before moving to quantum transport. This is what this book tries, self-consistently and briefly, to achieve.

Accordingly, the book is divided into 3 parts: Part I gives the necessary background in quantum mechanics and solid state physics pertaining to bands and velocities in semiconductors, i.e. the prerequisites for all types of transport. Part II gives the theory of classical transport and the theory of quantum conduction focusing mainly on the Landauer theory. The classical theory is immediately applied to the operation of the PN and Schottky junctions to make the connection between theory and applications transparent. Likewise, an analysis of the Resonant Tunnelling Diode is included in the chapter on quantum conduction as an immediate application of the Landauer theory. It is meant to serve the same connection. Part III deals with transistors. It begins with vacuum devices based on field emission usually omitted in devices textbooks. In my opinion they could not be left out in a book with "quantum" and "conduction" in its title. They also serve as a good stepping stone for the field effect transistor or FET. We also include an introduction to the recently invented vacuum FET. Then we analyze in detail the mechanisms of conduction—both classical and quantum—of the most prominent types of transistors, the Si MOSFET (long or nano-channel), the MODFET or HEMT, the quantum-well GaAs FET (QWFET), and also the still-under-investigation carbon nanotube FET (CNT-FET).

The book has been written with the final year and postgraduate electronics student in mind. Although it starts at an elementary level, it progressively gives all the material necessary to understand the research papers on devices published in the last decade and then be able to start research. Given, also, the detailed description (we believe) of most important physical concepts, we expect that the book will be of interest to materials science and physics students at the undergraduate level. I have assumed a minimum level of calculus. When 3-dimensional equations are needed, the 1-dimensional version is first derived. I have also assumed a familiarity with the existence of electrons and photons, though not holes, given in first-year undergraduate courses in the US and the UK. The two-slit experiment is also thoroughly described due to its weight in quantum mechanics.

I have tried to give a meticulous inspection of the text to avoid the usual misprints of any book, but if some remain the responsibility for these and the entire content of this book rests with the author.

About the Author

John P. Xanthakis earned his PhD from the Electrical Engineering Department of Imperial College, University of London in 1980. After a period as a post-doctoral fellow in the Mathematics Department of Imperial College, he obtained a lectureship at the National Technical University of Athens (NTUA) in 1985. In 1992–93 he spent the academic year at Imperial College on sabbatical leave. He was promoted to professor at NTUA in 1999. He has 50+ papers in peer-reviewed journals and numerous presentations at international conferences. He is a Senior Member of IEEE, a member of the New York Academy of Sciences, and a regular reviewer in many top journals. His main interest is in nanoelectronics, in particular, vacuum nanoelectronics.

I

Prerequisites: Quantum Mechanics and the Electronic States in Solids

CHAPTER 1

Quantum Mechanics

1.1 THE TWO-SLIT EXPERIMENT

The two-slit experiment plays a central role in quantum mechanics. As Feynman has said if someone does not understand the two-slit experiment he does not understand quantum mechanics. It is worth pointing out that it is a "gedanken" experiment, i.e. it exists only in our minds (for pedagogical purposes), but there are real diffraction experiments that are completely equivalent to it, so we always treat it as a real experiment.

The electron two-slit experiment is a replica of the electromagnetic (photonic) two-slit experiment originally performed by the English physicist Thomas Young in 1801. We describe it below. The apparatus involves a screen on which two pinholes (S_1, S_2) have been punctured. The screen is illuminated by a lamp emitting spherical waves of wavelength λ which fall onto the pinholes, see figure 1.1. Before we proceed further, it is important to clarify what is meant by "pinholes". If a is the width of the pinholes we must have $a \ll \lambda = $ *wavelength of light*, otherwise if $a \gg \lambda$ the features we are going to present will be absent. As the initial spherical wave falls onto the pinholes the latter become secondary sources and emit secondary spherical waves with their origin being the pinholes S_1, S_2.

Due to the design of the apparatus, the secondary sources are coherent which means that their frequency is the same and there is no time lag between them. Put simply, the photons from S_1, S_2 propagate in phase. As a consequence, any developed phase difference between the two secondary spherical waves can only come from a different path travelled by each. If the distances from their sources to the same point P on a second screen are r_1, r_2, then the path difference at P is $|r_2 - r_1|$, see figure 1.1 again. We remind the reader that a spherical wave has the form $r^{-1}\exp[i(kr - \omega t)]$.

Therefore, when the phase difference is an integer number of wavelengths there is an enhancement of waves and a bright spot appears on the screen at P. On the other hand, when the phase difference is an odd integer of half-wavelengths, there is a cancellation of waves and a dark spot appears on the screen. On the screen therefore an alteration of bright and dark fringes occurs which we will call interference fringes, see figure 1.2. An analysis found in all introductory optics books shows that the maxima are given by $y_m = mD\lambda/d$ where d is the slit spacing, D is the distance between the two screens and m is an integer.

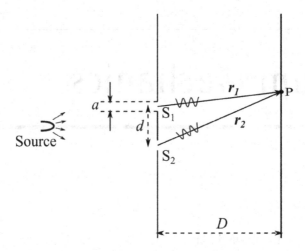

FIGURE 1.1 The two-slit experiment: two coherent waves from the slits S_1 and S_2 meet at a point P on a screen where they interfere constructively or destructively.

The exact algebraic relations which give the maxima and minima are not of real interest to us. The important point is that enhancement and cancellation can only occur with waves. Suppose we did not have a spherical wave in front of the two-slit screen but instead a series of pistols firing bullets (i.e. particles) to the screen at all angles and that the screen was impenetrable by the bullets. What type of distribution would we get on the second screen? We would get two gaussian-type distributions—shown in figure 1.2—each gaussian being positioned opposite each pinhole. When the experiment was performed by Davisson and Germer using electrons they did not get the bullet-type distribution but rather interference fringes, as shown in figure 1.2. Actually, as pointed out previously, they did not perform a two-slit experiment because they could not construct pinholes smaller than the wavelength of electrons but instead they performed an equivalent diffraction experiment of an electron beam off a metal surface. Therefore, electrons behave as waves. But what is it that

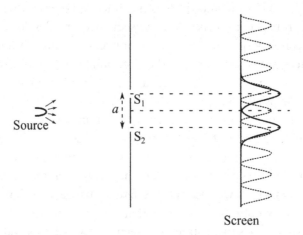

FIGURE 1.2 The expected pattern on the screen would be a set of interference fringes if the electrons are waves (dotted curve) or a set of two Gaussians if the electrons behave as particles.

is vibrating in electrons? We postpone the answer to this question for a moment to continue with the two-slit experiment.

Suppose we decrease the intensity of our electron gun (see later the field assisted electron emission phenomenon) so that only one electron passes at a time through the screen, i.e. perform the experiment as if electrons are fired by a gun. What do we see on the second screen? We still see the interference fringes but made out of dense spots. But we never see spots, i.e. electrons, on the minima of the fringes where the intensity was originally zero. Hence it seems that even a single electron can produce interference, that is a single electron can have wave properties while being a particle. All this is really mind blowing and the interpretation given by what is called the Copenhagen school of quantum mechanics—now accepted universally—is the following.

When electrons are measured they behave as particles but when they propagate in vacuum or in materials before measurement they behave as waves and so their position, upon measurement, is not predetermined but only probabilistically known. We must emphasize that when we talk about an electron wave it is wrong to think of the electron as a diffusive charge cloud. The correct picture or interpretation is that, due to the extremely small lengths involved, we can never tell exactly where it is, we can only talk about the probability of finding it in an infinitesimal space around a given position $\mathbf{r}$. It is this probability that behaves as a wave. Then it is the act of measurement that makes this probability certainty. In the two- slit experiment each electron gives two spherical probability waves $\Psi_1(r_1)$ and $\Psi_2(r_2)$—one from each pinhole—and the probability P of finding the electron in a small volume dV centered on any point given by its position vector $\mathbf{r}$ is

$$P(\mathbf{r}) = |\Psi(\mathbf{r})|^2 \, dV \qquad\qquad 1.1$$

where $\Psi = \Psi_1 + \Psi_2$

It must be clear from the discussion so far that the result of one position measurement can only be zero or one (0 or 1). Equation 1.1 becomes meaningful if many electrons are thrown onto the two-slit screen, i.e. if repeated measurements are made and then any positive value between 0 and 1 can occur. A more abstract interpretation of the experiment exists for which the reader can refer to advanced textbooks on Quantum Mechanics.

We can now come back to our original question of what it is that is vibrating in the electron wave. The answer is $\Psi(\mathbf{r})$ but only in the sense of equation 1.1. There is often an initial misunderstanding by some students that the electron is a "charge cloud". This is essentially wrong. In fact Schroedinger—whose equation we will discuss immediately below—originally thought that $|\Psi|^2$ was the charge density. Later developments by Born corrected that and completed the architecture of the Copenhagen school. On the other hand, $\Psi(\mathbf{r})$ should not be looked at as expressing only lack of information on behalf of the observer but rather as expressing a non-local nature of the electron.

The interference fringes are not produced by one electron going through one pinhole and another through the other pinhole. Remember the fringes are produced even when the electrons are thrown one by one onto the two – slit screen. If the two spherical waves emanating from the pinholes were not produced by the same electron they would not be coherent and interference fringes would not be produced. Therefore, the wavefunction Ψ refers to a single electron and not to a collection of many as a statistical average. The usual question asked in the lecture rooms is: how does the electron know that there is a second hole? The answer is that the electron is not a classical particle and it has a non-local character. In fact, non-locality is an inherent property of all particles. We will not delve any more into the particle–wave duality. We will accept on the one hand that electrons are countable and on the other we will deduce the properties of solids using the electronic wavefunctions Ψ.

1.2 THE SCHROEDINGER EQUATION, WAVEFUNCTIONS AND OPERATORS

It should be obvious from the previous discussion that we need to be able to evaluate $\Psi(r)$ in all cases. As an example of such a necessity we note that if we have the probability $P(r)$ corresponding to a wavefunction $\Psi(\mathbf{r})$, we automatically have the charge density $n(r)$ corresponding to this $\Psi(\mathbf{r})$ by the obvious relation

$$n(r) = eP(r) = e|\Psi(r)|^2 \qquad 1.2$$

We therefore seek an equation for $\Psi(r)$. This equation plays a central role in quantum mechanics and is called the Schroedinger equation after the famous Austrian physicist Erwin Schroedinger. The two-slit experiment can be interpreted by a wavefunction $\Psi(r) = \Psi_1(r_1) + \Psi_2(r_2)$, where the $\Psi_i(r_i)$ are spherical waves from the pinholes. Therefore, $\Psi(r)$ must be a solution of a wave equation in three dimensions. From previous courses we know that a wave equation must have the form

$$\nabla^2\Psi + k^2\Psi = 0 \qquad 1.3$$

where k has the dimensions of a wavenumber.

The value of k must conform with the principle of De Broglie, who associated with each particle of momentum p a matter wave of wavelength λ

$$\lambda = \frac{h}{p} \qquad 1.4$$

where h is Planck's constant. Equation 1.4 is usually rewritten in terms of the wavenumber k

$$k = p/\hbar \qquad 1.4a$$

with $\hbar$(pronounced h bar) $= h/2\pi =$ Planck's constant divided by 2π. Given that the energy of any particle can be analyzed into kinetic and potential energy V, i.e.

$$E = p^2/2m + V \qquad 1.5$$

we deduce that

$$k = p/\hbar = \frac{\sqrt{2m(E-V)}}{\hbar} \qquad\qquad 1.6$$

where m = mass of the particle. Then equation 1.3 takes the form

$$\frac{-\hbar^2\nabla^2}{2m}\Psi + V\Psi - E\Psi = 0 \qquad\qquad 1.7$$

Equation 1.7 is the time independent Schroedinger equation. We stress that we have not proven or derived anything. All the great equations of physics, like Newton's 3rd law or Maxwell's equations of electromagnetism, are ingenious generalizations and abstractions of observations and we have merely followed this course. In fact, we have a generalization: the two-slit experiment and the associated equation 1.3 pertain to electrons in vacuum. We have generalized it (or Schroedinger has in fact done so) to electrons in an arbitrary potential energy $V(x,y,z)$. The justification lies in the results. So far this equation has been used to successfully interpret myriad phenomena with no internal contradictions. A small caution on notation is necessary. In subsequent parts of the book where devices are analyzed we would be required to solve the Poisson equation for the potential which we will denote by V. Then the potential energy of an electron will be equal to (-eV). The reader will be warned of such a change in notation when it occurs.

We now rewrite equation 1.7 in the following form

$$\left[\frac{-\hbar^2\nabla^2}{2m} + V(r)\right]\Psi(r) = E\Psi(r) \qquad\qquad 1.8$$

If we treat

$$H = \frac{-\hbar^2\nabla^2}{2m} + V(r) \qquad\qquad 1.9$$

as an operator, equation 1.8 becomes

$$H\Psi_n = E_n\Psi_n \qquad\qquad 1.10$$

i.e. an eigenvalue equation where E_n are the eigenvalues and Ψ_n the eigenvectors. We have now reached a second deeper stage in our development of quantum mechanics. Firstly, the mathematical language of quantum mechanics is linear algebra. Every stationary, time independent, wavefunction $\Psi(r)$ is an eigenstate of the hamiltonian H. Secondly, there is a further generalization. The form of the hamiltonian H is such that it implies a relation between quantum mechanics and classical physics. If we look at this equation we observe that a) it has the dimensions of energy and b) the hamiltonian is derived from the classical expression for the energy (equation 1.5) if we substitute for the momentum p the operator $(-i\hbar\nabla)$, where $i = \sqrt{-1}$.

We are now ready to generalize our conceptual framework. This generalization is a big logical jump. It was originally an educated guess of ingenious physicists like Schroedinger and Heisenberg, but their ideas have stood the test of time. The rules of quantum mechanics that we are going to use (and it's only a subset of the complete set) are the following

i. Write down the classical expression for the energy, this is the classical hamiltonian.

ii. In this expression change all momenta p by the differential operator $(-i\hbar\nabla)$.

iii. The algebraic expression has now been turned into an operator, the hamiltonian operator. Compute its eigenvalues and eigenfunctions. Then the average value of repeated measurements of the position x of an electron with wavefunction $\Psi(r)$ is (remember $|\Psi|^2$ is probability)

$$\bar{x} = \int_{\text{all space}} x|\Psi(r)|^2 \, dV = \int_{\text{all space}} \Psi^*(r)x\Psi(r)dV \qquad 1.11$$

and in 3 dimensions (3D)

$$\bar{r} = \int_{\text{all space}} r|\Psi(r)|^2 \, dV = \int_{\text{all space}} \Psi^*(r)r\Psi(r)dV \qquad 1.11a$$

where in equation 1.11 V stands for volume. The average momentum (written directly in 3 dimensions) is

$$\bar{p} = \int_{\text{all space}} \Psi^*(\mathbf{r})(-ih\nabla)\Psi(r)dV \qquad 1.12$$

where in equation 1.12 we have written the grad in bold to make the 3-dimensional character of the equation explicit. We will not repeat this.

Since the square of $\Psi(\mathbf{r})$ is a probability we must have

$$\int_{\text{all space}} \Psi^*(\mathbf{r})\Psi(r)dV = 1 \qquad 1.13$$

i.e. the $\Psi(\mathbf{r})$ must be normalizable.

So far we have discussed stationary phenomena—the two-slit experiment is stationary as the fringes do not move with time. The Schroedinger equation discussed up to now is the time independent Schroedinger equation. But there is a time dependent equation. For the stationary systems we have discussed up to now, like that of the two-slit experiment, we have implicitly assumed that the wavefunctions must have a time dependent factor of the form $e^{-i\omega t} = e^{-i\left(\frac{E}{\hbar}\right)t}$. This is in full accordance with both the de Broglie hypothesis and Plank's principle that $E = \hbar\omega$. Note that this form is exactly the same as that of the stationary waves in electromagnetism and, most importantly, it gives a stationary charge density because

$$n(r) = eP(r) = e\Psi^*(r,t)\Psi(r,t) = e[\exp(i\omega t)\psi^*(r)\exp(-i\omega t)\psi(r)] = e\psi^*(r)\psi(r) \quad 1.14$$

where in 1.14 $\psi(r)$ is the space dependent part of $\Psi(r,t)$.

Differentiating such a non-stationary wavefunction we get

$$i\hbar\frac{\partial}{\partial t}\Psi = E\Psi$$

But $H\Psi = E\Psi$. Hence

$$i\hbar\frac{\partial\Psi}{\partial t} = H\Psi \qquad\qquad 1.15$$

This is the time dependent Schroedinger equation. Although we have derived it for non-stationary wavefunctions with a harmonic variation it holds for all type of time variations. We will not make extensive use of it in this book. We will mainly tackle stationary problems. With the exception of the important section 1.8, the symbol Ψ will denote a stationary wavefunction. Before moving to specific problems we note that the hamiltonian operator H belongs to the family of hermitian operators. These operators have real eigenvalues (an obvious requirement) and are defined by $H_{m,n} = H^*_{nm}$ where the symbol * denotes complex conjugate.

1.3 PARTICLE IN A RECTANGULAR BOX

There are only very few problems that can be solved analytically, that is without the help of a computer, just using the right mathematics. One of them that holds an archetypal position in quantum mechanics is the particle in a box. Traditionally it has been used to exemplify and make concrete all the quantum mechanical notions that we have presented so far. However, with the advent of nanoelectronics this simple problem has found practical applications apart from being an archetypal problem of quantum mechanics. We describe it in various approximations immediately below. We begin with the infinite depth quantum well in one dimension.

The well extends from $x = 0$ to $x = L$, (see figure 1.3), i.e. the potential energy is zero for $0 \leqslant x \leqslant L$ and infinite otherwise. We consider that an electron can not escape the well because it takes infinite energy to do so. Therefore, the boundary conditions of our problem are

$$\Psi(x) = 0 \text{ for } x = 0$$

$$\Psi(x) = 0 \text{ for } x = L$$

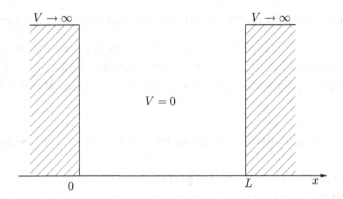

FIGURE 1.3   An infinite quantum well: a finite region where the potential V is zero inside, with infinite potential walls bounding it.

Inside the well, Schroedinger's equation in one dimension takes the form

$$-\frac{\hbar^2}{2m}\frac{d^2\Psi_n}{dx^2} = E_n\Psi_n \qquad\qquad 1.16$$

The general solution of this equation is

$$\Psi_n = A_n sin(k_n x) + B_n cos(k_n x) \qquad\qquad 1.17$$

For the boundary condition at $x = 0$ to hold, the B_n must be zero (because $cos(0) = 1$). Therefore

$$\Psi_n = A_n sin(k_n x) \qquad\qquad 1.17a$$

From the other condition at $x = L$ we get

$$sin(k_n L) = 0 \Rightarrow k_n = \frac{n\pi}{L}, n = 1,2,3\dots \qquad\qquad 1.18$$

We are left with only one set of unknowns, the A_n. Since Ψ is a probability we always have to satisfy the normalization condition, equation 1.13, so we have

$$\int_0^L \Psi^*\Psi dx = 1 \Rightarrow A_n = \frac{1}{\int_0^L sin^2(k_n x)dx}$$

After some very simple algebra we get

$$A_n = \sqrt{\frac{2}{L}} \qquad\qquad 1.19$$

If we substitute 1.17a–1.19 back into the Schroedinger equation, we get

$$E_n = \frac{n^2 \hbar^2 \pi^2}{2mL^2}$$

1.20

or using 1.18

$$E_n = \frac{\hbar^2 k_n^2}{2m}$$

1.21

The first three wavefunctions Ψ_1, Ψ_2, Ψ_3 together with their modulus squared are shown in figure 1.4. It can be seen that as n increases the nodes of the corresponding wavefunction Ψ_n increase accordingly. The electron in state n can be found anywhere with probability $|\Psi_n|^2$ but obviously never on the nodes. The most probable position for the nth state is

$$\overline{x_n} = \int_0^L \Psi^* x \Psi dx = \frac{L}{2}$$

1.22

which looks very plausible.

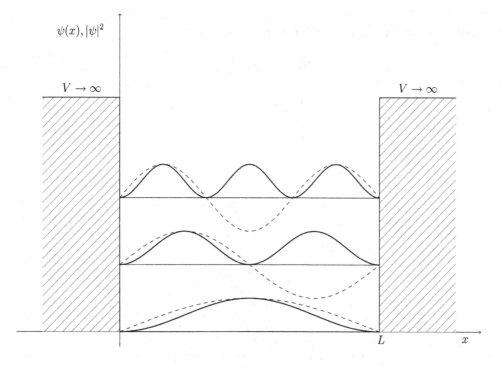

FIGURE 1.4   The first three eigenstates of an infinite quantum well.

As we can see from the results so far, instead of classical particles the electrons in a quantum well – box behave as oscillating strings i.e. waves and the greatest manifestation of their wave nature is in fact the discreteness of their energies.

The electron being in a zero potential energy state in the quantum box has all its energy as kinetic energy. From equation 1.21, we see that the expression $\hbar k_n$

$$\hbar k_n = \frac{n\pi}{L} = p_n$$

is the magnitude of the classical momentum of a particle. The reason for the above emphasis is that we are going to meet this expression again in vectorial form in the next section and in the chapter on the physics of electrons in solids. We now turn our attention to a generalization of the 1-dimensional quantum box to 3 dimensions.

The 3 – dimensional quantum box is shown schematically in figure 1.5. It extends along $0 \leqslant x \leqslant L_x$, $0 \leqslant y \leqslant L_y$, $0 \leqslant z \leqslant L_z$. Outside this region the potential is infinite so that the wavefunction Ψ is zero at the faces of the box.

The corresponding Schroedinger equation becomes

$$-\frac{\hbar^2}{2m}\nabla^2\Psi(r)=E\Psi(r)\Rightarrow \qquad\qquad 1.23$$

$$\Rightarrow \frac{\hbar^2}{2m}\left(\frac{\partial^2}{\partial x^2}+\frac{\partial^2}{\partial y^2}+\frac{\partial^2}{\partial z^2}\right)\Psi(r)+E\Psi(x,y,z)=0 \qquad\qquad 1.23a$$

If we assume a product form for the wavefunction

$$\Psi(r)=X(x)Y(y)Z(z) \qquad\qquad 1.24$$

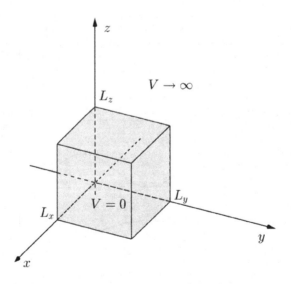

FIGURE 1.5 A 3-dimensional infinite quantum well.

then by the method of separation of variables we can decompose the 3-dimensional equation 1.23 into three 1-dimensional equations which have exactly the form of a 1-dimensional quantum well, i.e.

$$\frac{d^2X}{dx^2} + k_x^2 X = 0 \qquad\qquad 1.25a$$

$$\frac{d^2Y}{dy^2} + k_y^2 Y = 0 \qquad\qquad 1.25b$$

$$\frac{d^2Z}{dz^2} + k_z^2 Z = 0 \qquad\qquad 1.25c$$

where

$$k_x^2 = \frac{2mE^x}{\hbar^2} \qquad\qquad 1.26a$$

$$k_y^2 = \frac{2mE^y}{\hbar^2} \qquad\qquad 1.26b$$

$$k_z^2 = \frac{2mE^z}{\hbar^2} \qquad\qquad 1.26c$$

and E^x, E^y, E^z are the kinetic energies along the x, y, z directions respectively and $\hbar k_x$, etc., are the respective momenta in these directions.

Since the boundary conditions are the same as in the 1-dimensional case we can follow the same arguments for each direction that led us from equation 1.17 to 1.17a and then to 1.18 so that we finally get

$$\Psi_{n,m,l}(x,y,z) = C_{n,m,l} sin(k_x x) sin(k_y y) sin(k_z z) \qquad\qquad 1.27a$$

where

$$k_x = \frac{n\pi}{L_x}$$

$$k_y = \frac{m\pi}{L_y}$$

$$k_z = \frac{l\pi}{L_z} \qquad\qquad 1.27b$$

with m, n, l integers (positive or negative). There is no essential new physics here except the appearance of 3 quantum numbers which are associated with the three dimensions of space. We wish to emphasize that this is a general feature and we will always expect to find 3 quantum numbers in a realistic 3-dimensional physical problem such as any of the atoms of the periodic table. Every atom behaves as a 3-dimensional potential well, as we will see in section 1.6.

1.4 MORE QUANTUM MECHANICS, HEISENBERG'S UNCERTAINTY PRINCIPLE

Let us return to the 1-dimensional quantum well problem and calculate the average momentum according to equation 1.12. Application of this equation leads to (in 1-dimension)

$$\bar{p} = \int_0^L |A|^2 \sin(kx)\left(-i\hbar\frac{\partial}{\partial x}\right)\sin(kx)dx$$

$$\bar{p} = \int_0^L |A|^2 \sin(kx)(-i\hbar k)\cos(kx)dx = 0$$

since the integrand is an odd function of x. This result should not surprise us. In section 1.3 we found that the momentum is $\hbar k$ with $k = n\pi/L$ and with n taking both positive and negative integer values, so for each value of $\hbar k$ there is always its negative as a permitted value of the momentum . So if repeated measurements of momentum are made they can always be grouped as pairs of numbers of opposite sign and the average will give $\bar{p}$= zero.

This is an example of the strange two-fold rule of Quantum Mechanics. Prior to measurement of momentum of an electron with energy $\hbar^2 k^2/2m$ the electron behaves like a string, a wave, and one can only say that the momentum of the electron is either $\hbar k$ or $-\hbar k$. Upon measurement, it will only come out as one of them (but never zero) and exhibit particle behaviour. Prior to measurement we can only talk about a range of possible values, only after measurement do we get a definite value. We saw this principle in connection with the two-slit experiment and the measurement of position, but it holds for all physical variables. Prior to measurement the electron could be anywhere according to a probability given by the modulus of its wavefunction squared. After measuring it on a screen it has a definite position and behaves like a particle. The same principle holds for momentum and any physical variable in any system.

It is worthwhile discussing further the comparison of the quantum well to the free particle case. We saw in the first section of this chapter that a free electron in vacuum has a wavefunction of the form $e^{ik\cdot r}$. This wavefunction corresponds to a definite energy $E = \hbar^2 k^2/2m$ and a definite momentum $\hbar k$. One may wonder then why in the case of the quantum well with $V(x) = 0$ for $0 \leqslant x \leqslant L$ we have two values of momentum $\hbar k$ and $-\hbar k$. The answer is that $V(x)$ is not zero everywhere but only in the well. Outside the well the potential V is infinite which forces the electrons to bounce back at the well walls producing two running waves in opposite directions which in turn produce one standing wave as can be seen mathematically from the identity $2i\sin kx = e^{ikx} - e^{-ikx}$.

We now observe that the values $\hbar k$ are the eigenvalues of the momentum operator in one dimension. Indeed

$$-i\hbar \frac{\partial}{\partial x} e^{ikx} = -i\hbar \frac{\partial}{\partial x} e^{\frac{ipx}{\hbar}} = -(i\hbar)\left(\frac{ip}{\hbar}\right)e^{\frac{ipx}{\hbar}} = (p)e^{\frac{ipx}{\hbar}} \qquad 1.28$$

with e^{ikx} being the eigenfunctions. This can be extended to 3-dimensions

$$-i\hbar\nabla e^{\frac{ip.r}{\hbar}} = (p)e^{\frac{ip\cdot r}{\hbar}} \qquad 1.29$$

We now generalize this observation and state that the possible results of measurement of any physical quantity are the eigenvalues of an operator $\hat{O}(p,r)$ corresponding to this physical quantity. This operator is constructed by substituting for the classical momentum p the operator $(-i\hbar\nabla)$ and leaving the position vector r as it is, i.e. by the same rules as for the hamiltonian. Furthermore, $\hat{O}(p,r)$ is hermitian as is the hamiltonian. We will not need any other operators apart from the ones we have already introduced. However, it is helpful to know that this is a general property of quantum mechanics and it does not pertain to momentum only, although the latter forms the main part of the current density which is our main quantity of interest. These issues are further developed in Appendix A. The content of this Appendix is not necessary for the understanding of the rest of this book.

The above discussion of momentum brings us to the famous Heisenberg's uncertainty relation. We saw that in a 1-dimensional quantum well of size L, the most probable position of an electron is at the center, $x = \frac{L}{2}$ so that the uncertainty in its position is $\Delta x = L/2$. If we make a momentum measuremetnt at energy E_n (i.e. of a stationary state), we can get $\hbar k_n$ or $-\hbar k_m$, i.e. the average is zero and the deviation from the average = the uncertainty $\Delta p_n = \hbar n\pi / L$. If we multiply these two quantities together we get

$$\Delta x \Delta p > n\hbar\pi / 2 \qquad 1.30$$

Since the minimum value of the integer n is 1 we get

$$\Delta x \Delta p > \hbar \qquad 1.30a$$

This can be generalized for any physical system (not only for the 3-dimensional quantum well) to the following three relations

$$\Delta x \Delta p_x > \hbar \qquad 1.31a$$

$$\Delta y \Delta p_y > \hbar \qquad\qquad 1.31b$$

$$\Delta z \Delta p_z > \hbar \qquad\qquad 1.31c$$

The above relationships are not intuitive generalizations from the 3 dimensional rectangular quantum well but can be proved from the mathematical properties of the eigenvalues of linear hermitian operators. As the name suggests, it was Heisenberg who achieved that. Now, in analogy with the above relationships we also have the corresponding uncertainty relation for energy and time. We say "the corresponding" because the product of energy and time has the same physical dimension as the product of momentum and length, i.e. the dimension of action, that of the physical constant $\hbar$. Therefore, we have

$$\Delta E \Delta t \geqslant h \qquad\qquad 1.32$$

where Δt is the uncertainty of time (duration of measurement) and ΔE is the uncertainty in energy E. We will not occupy ourselves with this matter anymore as these relations will not be of direct use to us later.

A last point we would like to make is that in the free electron case the hamiltonian $p^2/2m$ and the momentum p commute as operators and hence they have common eigenfunctions, the e^{ikx}. When this happens in quantum mechanics for any two observables, the corresponding classical variables—i.e. the eigenvalues—can be measured simultaneously without error. This is the case of an electron in vacuum where both energy and momentum are known exactly.

1.5 STATISTICS OF ELECTRON OCCUPANCY, THE PAULI PRINCIPLE AND THE FERMI – DIRAC DISTRIBUTION

If we put N electrons in a quantum well, or in any system for that matter, how are they going to be distributed among the energy levels $E\left(k_x, k_y, k_z\right)$? The answer to this question will come in two stages: first by analyzing the so-called Pauli principle and then by analyzing the effect of the temperature T. We begin with the first. There are two kinds of particles in nature, fermions and bosons. Electrons are fermions which have the property that no two electrons in the same system can have the same quantum numbers. The quantum numbers in the 3-dimensional well are the k_x, k_y, k_z. In other systems, atoms for example, they may be different, but still each $\Psi(r)$ will be characterized by three indices corresponding to the three dimensions of the system.

These three indices are not enough to characterize the electron—a fourth related to spin is required. We assume that the reader is familiar with the concept of spin. If not, an introduction is given in Appendix A. For the purposes of this section it is enough to know the following: the electron is assumed to have an inherent magnetic moment $\mu_B = e\hbar/(2m)$, with m the mass of an electron which can point in nominally "up" or "down" directions. Therefore, given a state with quantum numbers k_x, k_y, k_z, we can put two electrons in it, one with "up" and the other with "down" spin. This does not violate the Pauli principle because they will differ in a fourth quantum number, a magnetic one which determines the direction of its magnetic moment. This is quantitatively discussed in Appendix A.

So at $T = 0$ it is easy to see what the distribution of the electrons among the energy levels will be—they will fill in ascending order (i.e. increasing energy) the levels $E(k_x, k_y, k_z)$ in pairs of spin "up" and "down" until $N/2$ levels have been filled up. If the system contains many electrons with their energy levels very close to each other, the highest occupied state is called "Fermi energy" and is usually symbolized as E_F.

But what is the effect of a finite temperature $T > 0$? We know that electrons, like any particle, can gain energy KT from the environment, (K is Boltzmann's constant), and jump to a higher empty or singly occupied state. Then the Pauli principle must still be observed. What is the rule for such transitions in an equilibrium state? It is given by the Fermi–Dirac probability of occupation of any state $P(E)$

$$P(E) = \frac{1}{1 + \exp\left(\dfrac{(E - E_F)}{KT}\right)} \qquad 1.33$$

where E_F does not necessarily have the previous meaning of the highest occupied energy level but it is just a variable to be determined by the total number of electrons, as we will find out.

The graph of the Fermi–Dirac distribution function is shown in figure 1.6. We observe that at $T = 0$ all the states with $E < E_F$ have $P(E) = 1$ and all the states with $E > E_F$ have $P(E) = 0$. So why does E_F does not have the meaning it has in the paragraph above? The answer is that E_F is not necessarily an allowed state. In metals, where there is a continuum of states, this distinction hardly makes any difference, but it does in semiconductors.

A careful examination of figure 1.6 shows that at $T > 0$ all the electrons that occupy the states with $E > E_F$ have come from the states a few KT below E_F. In fact the whole

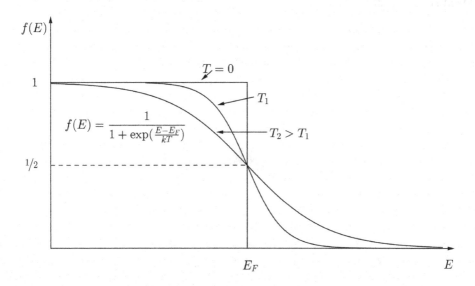

FIGURE 1.6 The Fermi–Dirac distribution function $f(E)$ which gives the probability of occupation of a state: $f(E)$ is 1 for $E < E_F$ and zero for $E > E_F$ at $T = 0$.

distribution $P(E)$ is unaltered at $E \ll E_F$ and $E \gg E_F$. All the changes have occurred a few KT below and above E_F. It is important to remember this as we shall see that all conduction mechanisms occur when electrons change state near E_F.

1.6 THE HYDROGEN ATOM AND THE ATOMS OF THE PERIODIC TABLE

We now have all the mathematical machinery to quantum mechanically analyze the simplest real physical system, the hydrogen atom. Before we do this, we examine two questions. What do we expect from the hydrogen atom based on classical mechanics? And why was quantum mechanics such a great triumph of human thought in the beginning of the 20th century? A simplified picture of a hydrogen atom is shown in figure 1.7. An electron of charge $-e$ at distance R is revolving around its nucleus of charge $+e$.

The coulombic force must act as a centrifugal force. Hence,

$$\frac{e^2}{4\pi\varepsilon_0 R^2} = mR\omega^2 \qquad\qquad 1.34$$

where $\omega = 2\pi f$ is the angular rate of rotation with f being the frequency. There is nothing in equation 1.34 to suggest that the radius R is quantized and consequently the energy is quantized. However, ample spectroscopic evidence existed before the discovery (or creation) of quantum mechanics to suggest that the energies were quantized. Spectroscopists found specific discrete frequencies coming out of the hydrogen atom and other atoms not a continuous range of frequencies, suggesting that discrete energy levels exist in the atoms. Furthermore, as is usually taught in advanced electromagnetism courses, an accelerating charge, like the electron in hydrogen, emits radiation so its orbit will not be circular but a spiral falling ultimately in the nucleus, something which clearly does not happen as hydrogen is a stable atom.

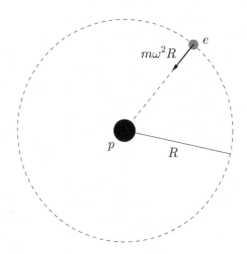

FIGURE 1.7   A classical picture of an electron (e) circulating around a proton (p) at a radius R with the centrifugal force $m\omega^2 R$ acting on it.

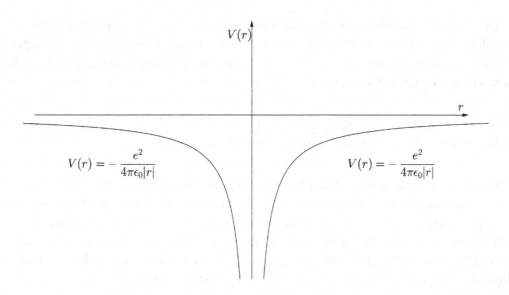

FIGURE 1.8   The Coulomb potential seen by an electron inside a hydrogen atom along two opposite directions.

We now begin our quantum mechanical analysis of atoms which will resolve all the above contradictions. We begin by writing down the classical equation of the energy of an electron in the field of the nucleus. The potential energy is due to the coulombic attraction that holds the electron close to the nucleus. The latter is shown in figure 1.8. Note that the radial distance r from the nucleus is always positive but what in figure 1.8 is meant by a negative r is the distance r in the opposite ($-180°$) direction from the one we have chosen to measure r initially. The reason we have chosen to present the coulombic attraction graphically this way is because it should be clear from figure 1.8 that the electron is again in a quantum well, albeit of a more complicated nature this time.

Now the electron in the hydrogen atom has kinetic and potential energy. Hence,

$$E = \boldsymbol{p}^2/2m + V(\boldsymbol{r}) \qquad\qquad 1.35$$

To construct the hamiltonian we have to apply the following rules of quantum mechanics: $E \to H$, $\boldsymbol{p} \to -i\hbar\nabla$ and $\boldsymbol{r} \to \boldsymbol{r}$.

Therefore, we get

$$H = \frac{-\hbar^2}{2m}\nabla^2 - \frac{e^2}{4\pi\varepsilon_0 r} \qquad\qquad 1.36a$$

for the hamiltonian and

$$H\Psi_i = E_i\Psi_i \qquad\qquad 1.36b$$

for the Schroedinger equation.

Unfortunately, this hamiltonian is in mixed cartesian and spherical coordinates. We therefore have either to express the operator ∇^2 in spherical coordinates or substitute $\sqrt{x^2 + y^2 + z^2}$ for r. The former choice is more convenient, so we proceed that way. Then the Schroedinger equation becomes

$$\left[\frac{-\hbar^2}{2m}\left(\frac{1}{r^2}\frac{\partial}{\partial r}\left(r^2\frac{\partial}{\partial r}\right) + \frac{1}{r^2 sin\theta}\frac{\partial}{\partial\theta}\left(sin\theta\frac{\partial}{\partial\theta}\right) + \frac{1}{r^2 sin^2\theta}\frac{\partial^2}{\partial\varphi^2}\right) - \frac{e^2}{4\pi\varepsilon_0 r}\right]\Psi(r,\theta,\varphi) = E\Psi(r,\theta,\varphi) \quad 1.37$$

This is an awkward looking partial differential equation that can, however, be decomposed into 3 separate ordinary differential equations just like we did in the 3-dimensional rectangular quantum box. The procedure is far from as straightforward as the simple rectangular well, so we do not find it useful to give this procedure here as it only entails pure mathematical steps of the separation of variables method. So we give the final results directly which have all the physical significance embedded in the equations. So for the wavefunctions we have

$$\Psi_{n,l,m_l}(r,\theta,\varphi) = R_{n,l}(r)Y_{l,m_l}(\theta,\varphi) \quad\quad\quad 1.38$$

We now analyze all the symbols in 1.38.

The n, l, m_l are quantum numbers labelling the wavefunctions just as in the 3-dimensional rectangular quantum well of the previous section we had 3 quantum numbers labelling the wavefunctions. The very simple reason why discrete states appear in the quantum version of the hydrogen atom and not in the classical is that the Schroedinger equation is an eigenvalue equation whereas the classical equations of motion do not have this property. All the mathematical steps we have omitted in going from equation 1.36 to the solution 1.38 for Ψ_{n,l,m_l} have this common theme. The quantum numbers n, l, m_l are given by the following formulae:

$$n = 1, 2, 3, 4, \ldots \quad\quad\quad 1.39a$$

$$l = 0, 1, \ldots, n-1 \quad\quad\quad 1.39b$$

$$m_l = -l, -l+1, \ldots, 0, ., +l \quad\quad\quad 1.39c$$

The eigenvalues corresponding to any Ψ are given by the equation

$$E_n = \frac{-me^4}{8\varepsilon_0^2 h^2 n^2} \quad\quad\quad 1.40$$

where h is Planck's constant (not h bar). As can be seen, the energies are negative because we have taken as the zero of energy the configuration where the electron is infinitely away from the nucleus, i.e. ionized. This has been implicitly assumed in the form of the coulombic potential in equation 1.36a. Furthermore, it is also evident that the eigenenergies depend only on the quantum number n and for this reason it is called the principal quantum number. Therefore, eigenfunctions with the same quantum number n but different l and m_l belong to the same energy. This is called degeneracy. We have already encountered this concept in connection with spin in discussing the Fermi–Dirac statistics. We remind the reader that the electron has an inherent spin or magnetic moment that can take only two values, "up" or "down". In the absence of a magnetic field, the "up" and "down" spin electrons have the same energy. Therefore, each Ψ_{n,l,m_l} can accommodate two electrons. Hydrogen has one electron in it, so that it will go to the lowest energy with $n = 1$. For this state, l can take only the value $l = 0$ and hence m_l can only be $m_l = 0$. The spin, however, is undetermined so we do not know in what spin state the electron is.

We move to the examination of the wavefunctions in greater detail. As can be seen from equation 1.38, the wavefunctions are a product of a function $R_{nl}(r)$, which is a function of only the radial distance r, and a function $Y_{lm}(\theta, \varphi)$ which is only a function of the angles θ, φ of the polar spherical system of coordinates. So the extent of the atom in space is determined by $R_{nl}(r)$. The function $Y_{lm}(\theta, \varphi)$—as a function of only the angles θ, φ—determines the distribution in space of $|\Psi|^2$. The Y_{lm} are called spherical harmonics. The Y_{lm} for various l and m_l are shown in figure 1.9. For the $l = 0$ as in the hydrogen case, they give a spherically symmetric distribution in space.

The R_{nl} in particular are a product of a polynomial $\mathcal{L}_{n,l}$ and a decaying exponential

$$R_{nl} = \mathcal{L}_{n,l}e^{-\beta r} \qquad\qquad 1.41$$

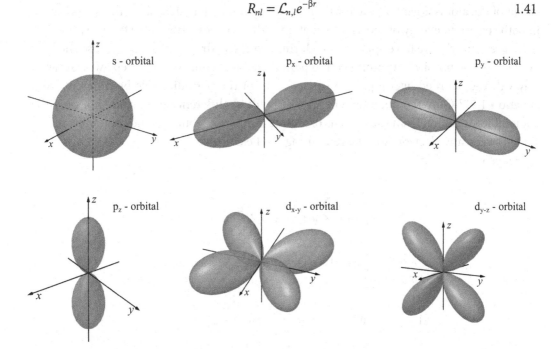

FIGURE 1.9   The spatial distribution of the Y_{lm} orbitals. For historical reasons the values for $l = 1, 3$ are denoted by the letters s, p, d respectively while the subscripts of s, p, d are notations for the third quantum number.

The units of β are obviously 1/*distance* so it is necessary at this point to define a characteristic length useful for atomic distances. The obvious way to do it is through equation 1.40 for the energy levels. We observe that E_n can be written as

$$E_n = -\frac{e^2}{8\pi\varepsilon_0 a_0}\frac{1}{n^2}$$ 1.42

where

$$a_0 \equiv \text{Bohr radius} \equiv \frac{4\pi\hbar^2\varepsilon_0}{me^2} = 0.529\text{Å}$$

The value of β can be written in terms of a_0. Calculations following the basic differential equation 1.37 (they are by no means short) show that

$$\beta = \frac{1}{na_0}$$ 1.43

Given that the extent in space of the hydrogen atom is determined by the decaying exponential in 1.41, we deduce the obvious result that the orbits (orbitals to be precise, as they only have a probabilistic nature) of states of higher energy lie further away from the nucleus. The extent of the atoms is generally a few Bohr radii. Equations 1.41–1.43 show the quantization in both space and energy of the hydrogen atom. Table 1.1 shows the $R_{nl}(r)$ for $n = 1, 2$.

The reason why we have spent so much time on the hydrogen atom is because the holy grail of atomic physics (the atoms of the periodic table) is only a small step away. In what way will the Schroedinger equation of any atom of the periodic table be different from equation 1.37? Only the potential energy $V(r)$ will be different and include not only the coulombic attraction from the nucleus of atomic number Z – but also the coulombic repulsion between any electron and the remaining $Z-1$ electrons.

Hence we can write

$$V(r) = -\frac{Ze^2}{4\pi\varepsilon_0 r} + \sum_{i=1}^{Z-1}\frac{e^2}{4\pi\varepsilon_0|r - r_i|} = -\frac{Ze^2}{4\pi\varepsilon_0 r} + V_{ee}(r)$$ 1.44

TABLE 1.1   Radial part of the hydrogen wavefunctions for n = 1 and n = 2

n	l	R_{nl}
1	0	$(1/a_0)^{3/2} \times 2exp(-r/a_0)$
2	0	$(1/(2a_0))^{3/2} \times \left(2 - \dfrac{r}{a_0}\right)exp(-r/2a_0)$
2	1	$(1/(2a_0))^{3/2} \times \left(\dfrac{r}{\sqrt{3}a_0}\right)exp(-r/2a_0)$

Obviously $V(r)$ in equation 1.44 will be a complicated expression of not only r but also of the r_i of the remaining electrons. However, if we decide that every electron moves in the average field of the other electrons and if we further assume that this is, to a first approximation, spherically symmetric, we can write

$$V(r) = \frac{Ze^2}{4\pi\varepsilon_0 r} + V_{ee}^{appr}(r)$$

1.45

where approximations $V_{ee}^{appr}(r)$ to $V_{ee}(r)$ can be found. How then would the wavefunctions of this hamiltonian look like? Answer: they are formally exactly the same as those of equation 1.38. The $R_{nl}(r)$ of course are not given by simple algebraic expressions as those of table 1.1 and can only be found numerically by solving by computer the corresponding differential equation for the $R_{nl}(r)$ but the angular part is given by the same functions, the spherical harmonics $Y_{lm}(\theta,\varphi)$. All these simplifications are a consequence of the spherical average we have taken for $V_{ee}(r)$.

As far as the eigenvalues are concerned those are not only a function of the first quantum number n but also a function of the second quantum number l, but not of the third one m_l, i.e. they have subscripts as E_{nl}. For historical reasons the values $l = 0,1,2,3$ are not given numerically but instead are designated by the letters s, p, d, f respectively. The electrons will fill in these states as described in section 1.5. However, to each eigenvalue E_{nl} we can have as many electrons as there are different m_l values and different spins. Hence the number of electrons in a given state E_{nl}, called an orbital, is therefore $N = 2(2l+1)$ where the 2 comes from spin and the $(2l+1)$ from the possible values of m_l. As we have stated before, this is the degeneracy of the atomic state or orbital. The atomic states are symbolized as, for example, $1s^1$ meaning the hydrogen state with $n=1$ and $l=0$ with one electron in it or $2p^4$ meaning a state with $n=2$ and $l=1$ with 4 electrons in it and so on. The atoms themselves are designated of course by the atomic number Z. The usual rule of high school chemistry, where a simple atom has to have 8 electrons in its outer shell to complete bonding, refers to atoms with s and p states in their outer shell. Note that 8 electrons are required to completely fill these states or orbitals. Figure 1.10 gives schematically the various atomic energies for an atom with a maximum quantum number $n=4$.

1.7 BARRIER PENETRATION, TUNNELLING

In section 1.3 we treated what we call bound states, i.e. states occurring in potential wells which are deep enough. As we saw, all atomic states are like that. We now come to discuss scattering states, i.e. propagating states which encounter a barrier in their way. The problem is linked to what is called tunnelling, i.e. the phenomenon of barrier penetration which is an entirely quantum mechanical phenomenon that frequently occurs in nanoelectronics. Imagine a mobile (a car, anything) of mass m travelling on a horizontal frictionless road with velocity v and encountering a bump

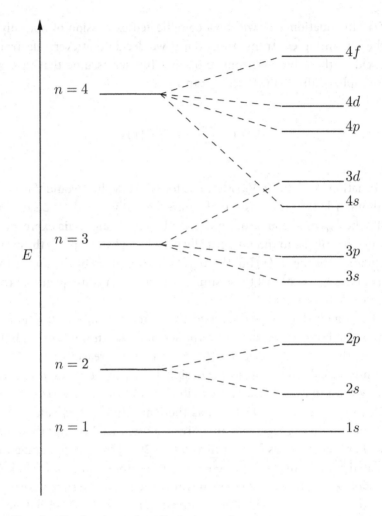

FIGURE 1.10   Notation for the eigenvalues $E_{n,l}$ in increasing energy.

of height h(see figure 1.11). The mobile will not go over the bump unless it has enough kinetic energy, in particular it will not go over the barrier unless

$$\frac{1}{2}mv^2 > mgh$$

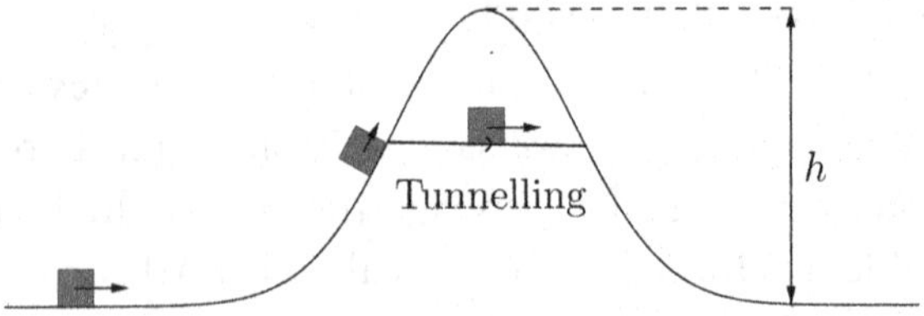

FIGURE 1.11   The concept of tunnelling: a classical mobile has to go necessarily over the barrier but a quantum particle can go through the barrier.

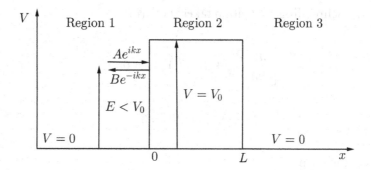

FIGURE 1.12   Incident and reflected plane waves on a rectangular barrier of height V_0.

where g is the gravitational constant. Surprisingly it will in quantum mechanics but with a small probability.

To analyze how this is possible we have to solve the Schroedinger equation again, but this time we have to assume an impinging wave to the barrier. So we assume free space everywhere except in the region $0 \leqslant x \leqslant L$ (see figure 1.12) where there is an electrostatic potential barrier of height V_o, equivalent to the gravitational barrier of figure 1.11. Onto this barrier impinges a wave e^{+ikx} travelling from left to right. If the energy E of the wave is less than V_o, classically we expect that all of the electrons in the wave will be reflected back, forming a wave of the form e^{-ikx}. But as we shall see, a very small portion of the incoming wave will go through.

So the Schroedinger equation in region 1 (see figure 1.12 again) takes the form (assuming $V = 0$ there)

$$\frac{-\hbar^2}{2m}\frac{\partial^2}{\partial x^2}\Psi_1 - E\Psi_1 = 0 \Rightarrow \Psi_1'' + k^2\Psi_1 = 0 \qquad 1.46$$

where $k^2 = 2mE/\hbar^2$ and the double upper dash symbol denotes double differentiation. The solutions to this equation are of the form

$$\Psi_1(x) = Ae^{ikx} + Be^{-ikx} \qquad 1.47$$

The critical reader can see that we don't have to assume an incoming wave as stated. Both incoming and reflected waves result from the Schroedinger equation. Of course, we have to assume that these states are filled, i.e. we have a beam of electrons. In connection with the Pauli principle, one may ask how come there can be at most 2 electrons per energy state and we have a beam which—as the word signifies—implies many electrons, all in the same energy E. The answer is that the energies of electron states in vacuum are very closely packed, they form a continuum, so that we can have many electron states in such a small energy interval ΔE, almost infinitesimal, so the Pauli principle is valid on the one hand and the electrons can be considered monoenergnetic on the other.

In region 2 the Schroedinger equation takes the form

$$\left[\frac{\hbar^2}{2m}\frac{\partial^2}{\partial x^2}+(E-V_o)\right]\Psi_2 = 0 \Rightarrow$$

$$\Rightarrow \Psi_2'' - a^2\Psi_2 = 0 \qquad\qquad 1.48$$

with solutions of the form

$$\Psi_2 = Ce^{ax} + De^{-ax} \qquad\qquad 1.49$$

In the remaining region 3, the Schroedinger equation takes exactly the same form as in region 1. So we expect the same solutions as in 1 but we know from physical intuition that only a small number of electrons will penetrate the barrier and there are no electrons coming from the right. (This will change when we encounter the Landauer formalism later). So we write

$$\Psi_3 = Ge^{+ikx} \qquad\qquad 1.50$$

We now come to the last ingredient of quantum mechanics: the wavefunction Ψ must be a smooth function of the coordinates. This means that both the function and its derivatives must be continuous. We may recall that the wavefunctions of the quantum well if examined in the entire range of $(-\infty, +\infty)$ do not have continuous derivatives at the well edges 0, L but this is only due to the fact that the well-depth is infinite, a highly unrealistic value. For all finite potential problems, the smoothness of the wavefunctions is a required property. Assuming continuity of Ψ and $d\Psi/dx$ at the boundaries 0, L we get

at $x = 0$

$$A + B = C + D \qquad\qquad 1.51a$$

$$ikA - ikB = aC - aD \qquad\qquad 1.51b$$

and at $x = L$

$$Ge^{ikL} = Ce^{aL} + De^{-aL} \qquad\qquad 1.51c$$

$$ikGe^{ikL} = aCe^{aL} - aDe^{-aL} \qquad\qquad 1.51d$$

Equations 1.51a–1.51d constitute a system of 4 equations with 5 unknowns that mathematically is indeterminate. However, the normalization condition provides the 5th equation.

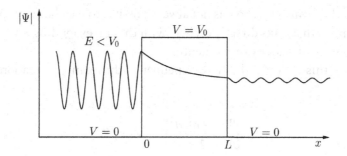

FIGURE 1.13 The solution of the Schroedinger equation for the rectangular barrier of figure 1.12; a travelling wave is found in the region to the right of the barrier.

Alternatively, we can set A=1. Then the calculated modulus of the wavefunction is shown in figure 1.13. We can see that this modulus decays exponentially inside the barrier but it does not go to zero. Hence a fraction of the incident electrons will be transmitted through the barrier. We only need the ratio of $(D/A)^2$ to determine this fraction of electrons. This ratio will also be the ratio of current densities measured in an experiment because the velocities in regions 1 and 3 are the same. The latter stems from the fact that the potentials there are the same. If they were not, we would have to multiply by the ratio of the velocities. Hence if we denote by T the ratio of transmitted to incident current and R the corresponding reflection coefficient

$$T = \frac{j_{inc}}{j_{tran}} = \left|\frac{G}{A}\right|^2 \qquad\qquad 1.52$$

$$R = 1 - T \qquad\qquad 1.53$$

The system of equations 1.51a–1.51d is a problem of straightforward but tedious algebra. An exact solution of B, C, D, and G can be found in terms of A, but we restrict the solution to the case where $aL \gg 1$ when an even simpler solution can be found. Then

$$T = 16\left(\frac{E}{V_o}\right)\left(1 - \frac{E}{V_o}\right)e^{-2aL} \qquad\qquad 1.54$$

The validity of this equation depends on the validity of the approximation $aL \gg 1$. Luckily this condition holds for many physical problems. We stress that equation 1.54 is only true for rectangular barriers. In the next section we obtain an approximate formula valid for most barriers of interest in nanoelectronics.

1.8 PROBABILITY CURRENT DENSITY AND THE WKB APPROXIMATION

So far we have implicitly obtained the current density J_i corresponding to a wavefunction Ψ_i as a product of $|\Psi_i|^2$ multiplied by a corresponding velocity. However, the velocity of an electron with a wavefunction Ψ_i and energy E_i is not always known unless Ψ_i is a free space wavefunction of the form $e^{\pm ipx/\hbar}$. So it is not always possible to use the above product. There is however a formula that links directly each Ψ_i with the corresponding J_i without the need to invoke the concept of velocity or momentum.

Let us see how this is done. The time dependent Schroedinger equation reads in one dimension

$$i\hbar\frac{\partial\Psi}{\partial t}=\frac{-\hbar^2}{2m}\frac{\partial^2\Psi}{\partial x^2}+V(x)\Psi \qquad\qquad 1.55a$$

Taking the complex conjugate

$$-i\hbar\frac{\partial\Psi^*}{\partial t}=\frac{-\hbar^2}{2m}\frac{\partial^2\Psi^*}{\partial x^2}+V(x)\Psi^* \qquad\qquad 1.55b$$

Now we focus our attention on a rectangular volume of space bound between the planes at points x_1, x_2(see figure 1.14), and we ask about the variation with time of the total probability in that volume.We initially assume only an x-dependence of the wavefunction $\Psi(x,t)$.

The probability of finding an electron in this volume of cross section A is

$$A\int_{x_1}^{x_2}\left|\Psi(x,t)\right|^2 dx = A\int_{x_1}^{x_2}\Psi^*(x,t)\Psi(x,t)dx \qquad\qquad 1.56$$

Its variation with time is given by

$$A\frac{\partial}{\partial t}\left[\int_{x_1}^{x_2}\Psi^*(x,t)\Psi(x,t)dx\right]=A\int_{x_1}^{x_2}\left[\frac{\partial\Psi^*}{\partial t}\Psi+\Psi^*\frac{\partial\Psi}{\partial t}dx\right]$$

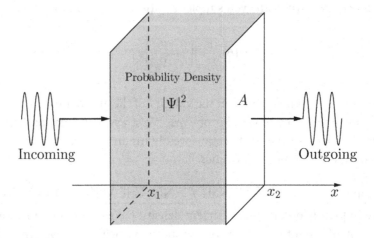

FIGURE 1.14   Geometry for the calculation of the time-dependence of probability density.

Using the time dependent Schroedinger equations 1.55a, 1.55b, we transform the right hand side (=RHS) of the above equation as

$$\frac{iA}{\hbar}\int_{x_1}^{x_2}\left[\left(\frac{-\hbar^2}{2m}\frac{\partial^2\Psi^*}{\partial x^2}+V\Psi^*\right)\Psi-\Psi^*\left(\frac{-\hbar^2}{2m}\frac{\partial^2\Psi}{\partial x^2}+V\Psi\right)dx\right]$$

Note the change of sign above from transferring the i from the denominator to the numerator. We therefore get by cancelling the $V\Psi$ terms and incorporating the cross section A as extra integrations

$$\frac{\partial}{\partial t}\int\int\int_{x_1}^{x_2}|\Psi|^2\,dxdydz=\frac{i\hbar}{2m}\int_{x_1}^{x_2}\left(\Psi^*\frac{\partial^2\Psi}{\partial x^2}-\Psi\frac{\partial^2\Psi^*}{\partial x^2}\right)dxdydz=$$

$$=\frac{i\hbar}{2m}\int_{x_1}^{x_2}\frac{\partial}{\partial x}\left(\Psi^*\frac{\partial\Psi}{\partial x}-\Psi\frac{\partial\Psi^*}{\partial x}\right)dxdydz \qquad 1.57$$

We come now to a 3-dimensional generalization of 1.57. We have written 1.57 in such a way, so that this generalization is straightforward. It's obvious that the 3-dimensional differential element $dxdydz$ is equal to the volume element dV and the 1-dimensional differential operator $\frac{\partial}{\partial x}$ will go to the grad operator ∇. Therefore, we get in 3-dimensions

$$\frac{\partial}{\partial t}\int_V|\Psi|^2\,dV=\frac{i\hbar}{2m}\int_V\nabla\left(\Psi^*\nabla\Psi-\Psi\nabla\Psi^*\right)dV \qquad 1.58$$

We now use the well-known formula transforming a volume integral to a surface integral

$$\int_V\nabla.\Phi dV=\int_S\Phi.dS$$

where S is the surface enclosing the volume V. We finally get

$$\frac{\partial}{\partial t}\int_V|\Psi|^2\,dV=\frac{i\hbar}{2m}\int_S\left(\Psi\nabla\Psi^*-\Psi^*\nabla\Psi\right)dS \qquad 1.59$$

What does equation 1.59 tell us? It tells us that the rate of change of probability density inside a volume V is equal to the flux of a quantity out of it. Therefore, that quantity must be the negative of the probability density current. Hence we obtain for the current that is due to a wavefunction Ψ_i

$$J_i(r,t) = \frac{-i\hbar}{2m}\left(\Psi_i\nabla\Psi_i^* - \Psi_i^*\nabla\Psi_i\right) \qquad 1.60a$$

For stationary states of energy E, the wavefunction will be of the form $\Psi(r,t) = \psi((r))e^{+iEt/\hbar}$ and the exponentials will cancel out, so that we can write

$$J_i(r) = \frac{-i\hbar}{2m}\left(\psi_i\nabla\psi_i^* - \psi_i^*\nabla\psi_i\right) \qquad 1.60b$$

If we apply this formula for example to the incoming to the barrier wave, we immediately get by elementary manipulations in 1-dimension

$$J = \frac{-i\hbar}{2m}(+ik+ik)A^*A = \frac{\hbar k}{m}A^*A = vA^*A \qquad 1.61$$

since $\hbar k/m$ is the velocity of the incoming electrons.

The main application of the formula 1.60b is not to derive again the expression for the current density in a simple case as the above but to derive an expression for the transmission coefficient of a general barrier and an expression for the current in the channel of a nanotransistor in later chapters of the book. For the time being we derive the required expression for the transmission coefficient of a general barrier without the resource of 1.60b and with a method not as rigorous as that based on 1.60b but quicker and more physically transparent. We will revisit this problem more rigourously in a later chapter.

Consider then a 1-dimensional general barrier characterized by the potential function $V(x)$, as in figure 1.15, and an incoming electron of energy $E < V(x)$ for all x. The barrier can be divided into thin rectangular slices as shown again in figure 1.15. Now the wavefunction of the electron must look like e^{ikx} far to the left and far to the right of the barrier so the problem is in this respect similar to the problem of section 1.7. We can imagine that this electron wavefunction goes through the succession of rectangular barriers into which the general barrier can be divided and each time it is attenuated by the transmission coefficient of the ith rectangular barrier i.e. by

$$T_i \approx exp\left(-\frac{2\sqrt{2m(V_i - E)}\Delta x_i}{\hbar}\right) \qquad 1.62$$

where V_i is the local barrier height of the ith slice and Δx_i is its width. We have omitted the preexponential factor as this is usually very small compared to the exponential. We also assume that there is no reflection at each of the individual slices of the barrier.

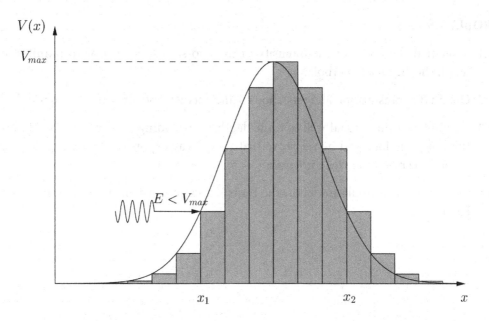

FIGURE 1.15   A wave incident on an arbitrary barrier can be thought to be transmitted through the rectangular layers into which the barrier can be divided.

The total transmission coefficient T will be the product of all the T_i, that is

$$T = T_1 T_2 \ldots T_N \qquad\qquad 1.63$$

where N is the total number of slices into which the barrier has been divided. Therefore

$$T = exp\left[-\sum_{i=1}^{N} \frac{2\sqrt{2m(V_i - E)}}{\hbar} \Delta x_i \right] \qquad\qquad 1.64$$

But in the limit of very large N the summation is simply an integral. Hence

$$T = exp\left[-2\int_{x_1}^{x_2} \frac{\sqrt{2m(V(x) - E)}}{\hbar} dx \right] \qquad\qquad 1.65$$

where x_1, x_2 are the points that define the barrier, i.e. x_1, x_2, are such that in the interval $[x_1, x_2]$ we have $E < V(x)$. Formula 1.64 can be obtained by a more polished version of 1.60, that is, by an approximation called the WKB approximation from the names of the authors of the original papers. We will prove this approximation later in the text and use it in both the theory of electron emission and the nanotransistor, but it will be sufficient for now to say that it is strictly valid for 1-dimensional problems and it is only valid for "deep tunnelling", i.e. when the energy of the electron E is well below the barrier maximum.

PROBLEMS

1.1 Show that the nodes of the intensity in the two-slit experiment with infinitesimal width slits lie on a hyperbola.

1.2 Obtain the transmission and reflection coefficients at a potential step of magnitude V_0.

1.3 Assume a 1-dimensional well of finite depth V_0 extending from $x = 0$ to $x = L$. Find the wavefunctions and hence prove that they decay exponentially outside the well. Extend the result to three dimensions.

1.4 Use the WKB formula to obtain the transmission coefficient through a trapezoidal barrier.

CHAPTER 2

Electron States in Solids

2.1 QUALITATIVE DESCRIPTION OF SOLIDS AND THEIR ENERGY BANDS

Now that we have developed our quantum mechanical weapons and have a proper theory of atoms, we can move to the description of energy bands in solids, which are the building blocks of micro- and nano-devices. Before we embark into a quantitative analysis, we first give a qualitative description of the formation of energy bands in solids.

Hydrogen is a gas: when two hydrogen atoms come near each other, they form a diatomic gas molecule. But for the purpose of exploring the simplest possible example showing the mechanism behind the formation of any solid from its atoms, imagine the following thought experiment: we have a collection of N hydrogen atoms as an 1-dimensional chain spaced a mile apart from each other. Every electron in each hydrogen atom will see only the attractive potential of its own nucleus and therefore it will be in the 1s state as described in the previous chapter and it will have the same 1s energy of $13.6eV$, see figure 2.1.

Now imagine a godly hand pushing the atoms towards each other along the chain direction. Let the interatomic distance be denoted by d (a variable) and construct a diagram of their energies E as a function of d. Initially all the energies of the N electrons (in 1s state) will be at the same level E equal to $13.6eV$. But as the atoms approach each other, their electrons will not only see the coulombic attraction of their own nucleus, but also the attraction of the neighbouring nuclei, as well as the repulsion of neighbouring electrons until bonds are formed and d attains a final value at $d = a$. Then, because of the new coulombic interactions, the electron 1s energies will no longer be equal to each other, as they initially were, but will split into N different values and then a band is formed, see figure 2.2. Pauli's principle prevents the states from being filled by more than 2 electrons (with different spins) so that the end result can be described as follows.

We have N atoms and N electrons in 1s states so that we have N states for spin up and N states for spin down, $2N$ in total. The N electrons will occupy the lowest N states and the remaining N higher states will be empty. As we will see later in the chapter, this is the picture of a metal. If hydrogen atoms had formed a solid, it would have been a metal. The critical reader will notice that we say "states" instead of atomic orbitals. There is a reason

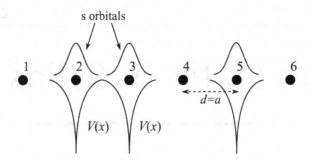

FIGURE 2.1   A linear chain of N hydrogen atoms, each having an electron in an $1s$ state. They are at a variable distance d, which attains a final value a when bands are formed.

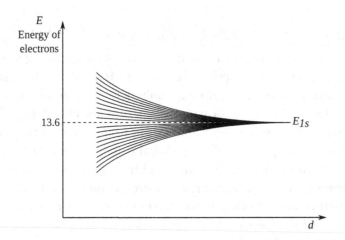

FIGURE 2.2   The splitting of the original N $1s$ energies – all equal to $E_{1s} = 13.6eV$ – into a band.

for it: although the electrons started as atomic orbitals when they were far apart, they are no longer in atomic orbitals when the atoms end up close together at Angstrom distances in a solid.

The purpose of this chapter is to explore the type of states we have in a solid and set up the ground for their conduction properties. However even before we move on to the next chapter, with just the knowledge of this chapter, we will be able to answer such questions as:

a. Why Al is a metal while GaAs is a semiconductor

b. Why intrinsic Ge has more carriers (electrons in its conduction band) than intrinsic Si

c. Why GaAs has a higher conductivity than Si given the same amount of doping

2.2   THE k-SPACE, BLOCH'S THEOREM AND BRILLOUIN ZONES

In nature there are two types of solids: the crystalline and the amorphous. We will be dealing with only the crystalline phase because the vast majority of electronic devices are made from crystalline materials. There are more applications for crystalline materials than

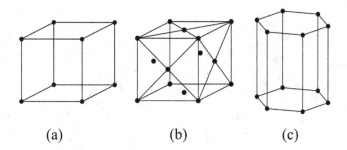

(a)   (b)   (c)

FIGURE 2.3 Three examples of Bravais lattices: the cubic (a), the face-centred (b), and the hexagonal (c).

for amorphous materials. Crystalline materials are those in which the atoms inside them are arranged periodically. Three particular examples, the simple cubic, the face-centred cubic, and the hexagonal, are shown in figure 2.3a-c. The vast majority of semiconductors belong to the face-centred cubic type, usually abbreviated as FCC. Now we must differentiate between what is called a Bravais lattice and the real lattice of atoms. A Bravais lattice is one whose points R_i are given by the relation

$$R_i = m_i^1 a_1 + m_i^2 a_2 + m_i^3 a_3 \qquad\qquad 2.1$$

where m_i^1, m_i^2, m_i^3 are integers and the a_1, a_2, a_3 are unit vectors in three dimensions, not necessarily orthogonal to each other. Figure 2.4 illustrates this non-orthogonality in two dimensions for the hexagonal lattice. There are 14 types of such lattices. Obviously equation 2.1 defines a periodic arrangement of the R_i. So in what way is a real lattice of atoms different from a Bravais lattice? A real lattice coincides with a Bravais lattice if the mathematical points defined by 2.1 are each occupied by one atom.

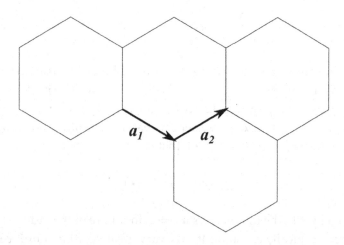

FIGURE 2.4 The planar hexagonal lattice with an non-orthogonal basis (a_1, a_2).

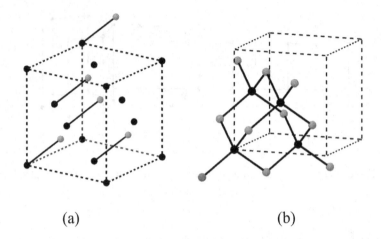

(a)   (b)

FIGURE 2.5   The zinc-blende lattice is a FCC Bravais lattice with each Bravais lattice point occupied by a diatomic unit as shown selectively in (a). The resulting structure has each atom surrounded by 4 others of the opposite type, in a tetrahedral arrangement, as shown again selectively in (b).

However, each mathematical point defined by 2.1 may not be occupied by one atom but by a many-atom molecular unit. Then we have the real lattice of a solid substance. In GaAs, for example, each Bravais point is occupied by one Ga and one As atom. Hence the real solids are composed by as many interpenetrating Bravais lattices as the number of atoms in their molecular unit. In figure 2.5a, we show the FCC lattice of the As atoms (black) and selectively the Ga atoms that each accompany an As atom. The end result is the so-called zinc blende or diamond lattice that is shown in figure 2.5b. The volume assigned to this molecular unit, which if repeated according to 2.1 fills the entire space, is called the unit cell. Often in the literature the unit cell may contain more than one unit. This is done for a better pictorial representation.

The Schroedinger equation for the crystalline solid reads

$$\left[\frac{-\hbar^2\nabla^2}{2m}+V_{cr}(r)\right]\Psi_i(r)=E_i\Psi_i(r)$$

where the crystalline potential energy $V_{cr}(r)$ includes all the electrostatic interactions of a given electron with all the remaining ones and with all the positive nuclei. Now, given that the Bravais lattice underlying any real lattice of atoms has the same molecular unit on each of its points and given that is periodic, we deduce that the crystalline potential energy $V_{cr}(r)$ must also be periodic with the same period, that is

$$V_{cr}(r+R_i)=V_{cr}(r) \qquad\qquad 2.2$$

for all the R_i given by 2.1. The unit vectors a_j, $j=1,3$ in 2.1 are of the order of Angstroms so that $V_{cr}(r)$ is macroscopically constant. It may vary greatly within a unit cell but it is the same from unit cell to unit cell, so we expect the wavefunctions not to differ much from the plane waves of vacuum where the potential is constant. (Whether the potential is constant or zero in space makes little difference since any constant potential can be made zero by a proper choice of energy).

In fact, Bloch's theorem says that the wavefunctions are of the form

$$\Psi_k(r) = U_k(r)e^{ik\cdot r} \qquad\qquad 2.3$$

where k is a wavevector in 3-dimensions and $U_k(r)$ is the same in each unit cell, i.e. the $\Psi_k(r)$ are modulated plane waves. Figure 2.6 illustrates how $\Psi_k(r)$ and its components vary in space. It should now be obvious that the wavefunctions in a solid are not simple atomic orbitals but wavefunctions that extend from one end of the solid to the other, something we hinted at in the first section of this chapter. However, the wavefunctions in a solid can be constructed from the atomic orbitals of their constituent atoms and then the connection between the simplified picture of section 2.1 and k space will become evident in section 2.3. The drawing of V(x) in Figure 2.1 gives an adequate 1-dimensional picture of the behaviour of the crystalline potential.

We will initially give a simple 1-dimensional proof of Bloch's theorem. However, this simple proof contains all the physical insight that lies behind the 3-dimensional proof that requires group theory to be complete. Let us imagine a linear chain of atoms like that of figure 2.1 consisting of N atoms with a distance a between the atoms and of total length L. Periodic boundary conditions on the chain are assumed, i.e. the wavefunction is the same

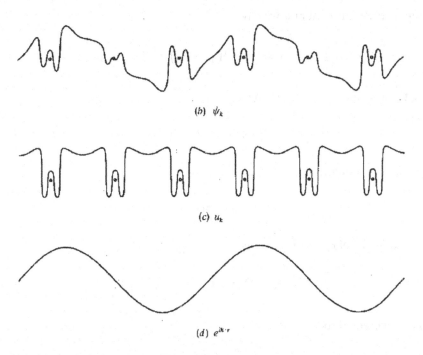

(b) ψ_k

(c) u_k

(d) $e^{ik\cdot r}$

FIGURE 2.6 The total (b), cellular (c), and plane wave (d) parts of the wavefunction Ψ. Figure reproduced from the book "Solid State Theory" by W. Harrison, Dover publications, 1980.

at the end atoms. This can be achieved by assuming that the chain is bent on itself and becomes a ring.

A more formal statement of the periodic boundary condition is

$$\Psi(x) = \Psi(x + L) \qquad 2.4$$

where (assuming L is very large) $L = Na$. Now since the potential energy is periodic with period a, the charge density must also be periodic, that is

$$\rho(x) = \rho(x + a)$$

But given that the relation between the charge density of a single electron $\rho'(x)$ and its wavefunction is $\rho'(x) = e|\Psi(x)|^2$, the above equation for the charge density of the whole crystal can only be true if for every wavefunction we have

$$|\Psi(x + a)|^2 = |\Psi(x)|^2 \qquad 2.5$$

When the magnitude of two complex numbers is the same, the complex numbers $\Psi(x)$, $\Psi(x+a)$ can only differ by a phase factor. We then get

$$\Psi(x + a) = e^{i\theta}\Psi(x) = \lambda\Psi(x) \qquad 2.6$$

We can repeat the argument and we have

$$|\Psi(x + 2a)|^2 = |\Psi(x + a)|^2 \Rightarrow \Psi(x + 2a) = \lambda\Psi(x + a) = \lambda^2\Psi(x)$$

Repeating the argument N times we have

$$\Psi(x + Na) = \lambda^N\Psi(x) \qquad 2.7$$

By comparing now equations 2.4 and 2.7 we deduce

$$\lambda^N = 1$$

Hence λ must be the N roots of 1, that is

$$\lambda = e^{2\pi i v/N} \quad (v = 0, 1, 2, \ldots, N-1) \qquad 2.8$$

so that θ in 2.6 is equal to

$$\theta = \frac{2\pi v}{N} = \frac{2\pi v a}{Na} = \frac{2\pi v}{L}a \qquad 2.9$$

The term $2\pi v/L$ has the dimension of a wavevector (1/distance), so that we can finally write

$$\Psi(x+a)=e^{ika}\Psi(x)$$

where $k = 2\pi v/L, v = 0,1,2,\ldots,N-1$ and more generally

$$\Psi(x+na)=e^{ikna}\Psi(x) \qquad\qquad 2.10$$

We have to generalize equation 2.10 to 2- and 3-dimensions. Before giving the proof in 2- and 3-dimensions we can do a bit of guesswork as follows: na is the R_n position vector in this linear 1-dimensional chain, so that the phase factor in front of $\Psi(x)$ can be written e^{ikR_n}. To guess correctly the desired result in 3-dimensions, we expect that we only need to

1. treat R as a vector $\boldsymbol{R}$,

2. turn the scalar x into the position vector r,

3. turn the simple scalar wavevector k into the vector $\boldsymbol{k}$, which also labels the wavefunction $\Psi_k(r)$ and finally

4. turn the simple product into a dot product

Then we expect to obtain

$$\Psi_k(r+\boldsymbol{R}_j)=e^{i\boldsymbol{k}\cdot\boldsymbol{R}_j}\Psi_k(r) \qquad\qquad 2.11$$

Equation 2.11 is indeed the correct relation and is called the Bloch condition. It is very easy to verify that equations 2.3 and 2.11 are equivalent. Simply substitute $r+\boldsymbol{R}_n$ for r in 2.3 and you will immediately get 2.11 given that $U_k(r)$ is periodic. Although we correctly guessed the 3-dimensional generalization of 2.10, this guesswork has not helped us to specify the 3-dimensional wavevector $\boldsymbol{k}$. We will now follow a more formal path in 2 dimensions which can be easily generalized to 3 dimensions.

Let us first rewrite the 1-dimensional k as

$$k=\frac{2\pi v}{L}=\frac{2\pi v}{Na}=\frac{2\pi v}{N}\cdot\frac{1}{a} \qquad\qquad 2.12$$

Now imagine a 2-dimensional rectangular lattice in the x and y directions with spacings a_1, a_2 between the atoms in the respective x, y directions as in figure 2.7. Let us also assume periodic boundary conditions along both the x and y directions as we did for the linear

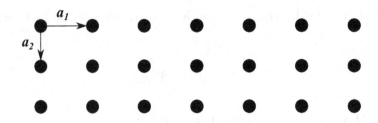

FIGURE 2.7   A square lattice with an orthogonal basis (a_1, a_2).

chain. The series of arguments leading from 2.5 to 2.10 can be repeated here along each of the x and y directions separately leading to

$$\Psi(x+na_1, y+ma_2) = e^{i(k_x na_1 + k_y ma_2)}\Psi(x, y)$$ 2.13

or in vector form

$$\Psi(r+R_j) = e^{ik\cdot R_j}\Psi(r)$$ 2.14

where k is now a 2-dimensional vector given by

$$k = 2\pi\left(\frac{v_1}{N_1}\frac{1}{a_1}, \frac{v_2}{N_2}\frac{1}{a_2}\right)$$ 2.15

with N_1, N_2 the number of unit cells in the x and y directions and $v_1 = 0,1,2,\ldots,N_1-1$ and $v_2 = 0,1,2,\ldots,N_2-1$. The above equation may be written

$$k = 2\pi\left(\frac{v_1}{N_1}b_1 + \frac{v_2}{N_2}b_2\right)$$ 2.15a

where b_1 is a unit vector along the x direction and of magnitude $\frac{1}{a_1}$ and b_2 is a unit vector along the y direction and of magnitude $\frac{1}{a_2}$. Equation 2.14 is identical to equation 2.11, so one would have thought that we have proven Bloch's theorem in at least 2 dimensions. However, we have not, the reason being that in our proof we have used a rectangular lattice, which constitutes a simplification in that the unit vectors of such a lattice are at right angles to each other. However, as can be seen from figure 2.4, all lattices do not possess an orthogonal set of unit vectors.

The general proof is, as we stated earlier, a bit more complicated and we will not give it here. However, the simple proof has all the physical insight that we need so that the general case can be made transparent. The general case can be stated as follows. Let a_1, a_2, a_3 be the unit vectors of the 3-dimensional Bravais lattice of the solid in question. Define the so called reciprocal lattice unit vectors by

$$\boldsymbol{b}_1 = \frac{\boldsymbol{a}_2 \times \boldsymbol{a}_3}{\boldsymbol{a}_1 \cdot \boldsymbol{a}_2 \times \boldsymbol{a}_3}$$

$$\boldsymbol{b}_2 = \frac{\boldsymbol{a}_3 \times \boldsymbol{a}_1}{\boldsymbol{a}_1 \cdot \boldsymbol{a}_2 \times \boldsymbol{a}_3}$$

$$\boldsymbol{b}_3 = \frac{\boldsymbol{a}_1 \times \boldsymbol{a}_2}{\boldsymbol{a}_1 \cdot \boldsymbol{a}_2 \times \boldsymbol{a}_3} \qquad 2.16$$

Then the $\boldsymbol{k}$ vectors consistent with or obeying Bloch's theorem (equation 2.11) are

$$\boldsymbol{k} = \frac{2\pi v_1}{N_1}\boldsymbol{b}_1 + \frac{2\pi v_2}{N_2}\boldsymbol{b}_2 + \frac{2\pi v_3}{N_3}\boldsymbol{b}_3 \qquad 2.17$$

This is the generalization of equation 2.15 for any of the 14 Bravais lattices where N_i, i=1,3 (meaning 1 to 3) are the number of unit cells of the Bravais lattice in the ith direction and $v_i = 0,1,2,\ldots,N_i$-1. Note that for a rectangular (or a cubic) lattice, where the $\boldsymbol{b}_i$ are orthogonal to each other, equation 2.15 forms the 2D part of 2.17.

Equations 2.16 and 2.17 have a simple geometrical interpretation. Just as the three $\boldsymbol{a}_i$, i=1,3 define a lattice of isolated points in real $\mathbf{r}$ space in 3 dimensions, so do the three $\boldsymbol{b}_i$, i=1,3 in $\boldsymbol{k}$ space. This lattice is called the reciprocal lattice The relation between the $\boldsymbol{a}_i$ and $\boldsymbol{b}_i$ is

$$\boldsymbol{a}_i . \boldsymbol{b}_i = 2\pi\delta_{ij} \qquad 2.18$$

The reciprocal lattice points $\boldsymbol{K}_i$ are obviously defined by the equation

$$\boldsymbol{K}_i = n_i^1 \boldsymbol{b}_1 + n_i^2 \boldsymbol{b}_2 + n_i^3 \boldsymbol{b}_3$$

where the n_i^j, j=1,3 are integers. In fact, equations 2.16 constitute the solution of 2.18 for the $\boldsymbol{b}_i$. Then the $\boldsymbol{k}$ vectors which are given by equation 2.17 and which label the Ψ_k are all contained in the unit cell of this reciprocal lattice. The inverse lattice of a cubic lattice of size a is also a cubic lattice of size $1/a$, the inverse of an FCC lattice of unit length a is a BCC lattice of unit length $1/a 1/a$ and likewise that of a BCC lattice is an FCC, hence the name reciprocal or inverse.

So given the direct lattice, the inverse lattice (which amounts to computing the $\boldsymbol{b}_j$ in 2.16) is a lattice as equally simple as the direct. At any rate, the inverse lattices for all the 14 Bravais lattices have been computed and are known but, as has been stated before, most semiconductors relevant to electronics like Si, Ge, and GaAs crystallize in the FCC lattice and

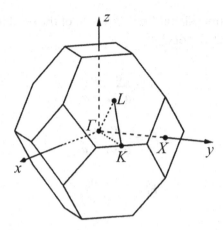

FIGURE 2.8 The Brillouin zone of the FCC lattice.

have a BCC inverse lattice. Figure 2.8 portrays the unit cell of the reciprocal lattice of the FCC lattice. The latter is called Brillouin zone (or first Brillouin zone) and this name stands for the unit cell of all reciprocal lattices, not just for the FCC. The number of k vectors in the first Brillouin zone is enormous: it is exactly $N_1 \times N_2 \times N_3$ which is of the order of Avocadro's number=10^{23}/mole. We can therefore treat k as a continuous variable.

A brief summary is worthwhile before closing this section.

1. A crystalline solid is a solid with a periodic arrangement of the atoms

2. The crystalline lattice of a crystalline solid has an underlying lattice called a Bravais lattice. The Bravais lattice is defined as the set of points on each of which a molecular unit of the solid "sits" and, if it is repeated according to the translations of equation 2.1, it reproduces the whole crystal

3. The periodicity of the crystalline potential imposes a condition on the allowed wavefunctions, the Bloch condition, i.e. equation 2.11 which we repeat here

$$\Psi_k\left(r+R_j\right)=e^{ik\cdot R_j}\Psi_k\left(r\right)$$

Consequently, the $\Psi_k(r)$ are waves that fill the entire crystal

4. The k vectors labeling the wavefunctions are defined as follows: given the Bravais lattice of the solid, the reciprocal lattice is obtained using the unit vectors defined by 2.16. Then the allowed k are all the vectors (defined by 2.17) that lie in the unit cell of the reciprocal lattice, which is called the "Brillouin zone"

2.3 THE LCAO METHOD OF CALCULATING ENERGY LEVELS

We now come to piece together the knowledge that we have previously obtained and see how metals, semiconductors, and insulators arise from common principles. This will be achieved by the linear combination of atomic orbitals method (abbreviated LCAO). This

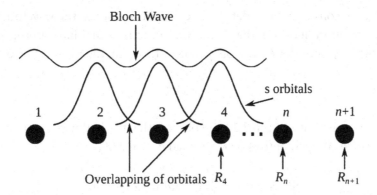

FIGURE 2.9 The schematic formation of Bloch waves in the linear chain of hydrogen atoms according to the LCAO method.

method can be used as both a first principles method or a semiempirical method, a distinction which will become evident later.

As usual, we will begin with a simplified model. Consider the linear chain of atoms described in section 2.1. We redraw it here in figure 2.9, showing the 1s orbital of each hydrogen atom when the atoms are near enough to have formed a solid. Our aim is to see how the Bloch functions are formed out of the original orbitals of the isolated atoms. We assume that the linear chain has been stabilized at an interatomic distance found in most solids, i.e. that of $2-3A^o$. As a consequence, the overlapping between a 1s orbital and a far distant one must be negligible but that between nearest neighbour atoms is significant. Futhermore, as a simplification we will assume that with second and further nearest neighbours overlapping is zero, as shown in figure 2.9. This is only approximately true—overlapping of atomic orbitals may occur up to third nearest neighbours, but their overlapping (defined mathematically below) decays rapidly with distance, so our approximation will not qualitatively change our results. Mathematically our approximation can be written as

$$\int \Phi^*(x - R_n)\Phi(x - R_m)dx = 0 \qquad 2.19$$

if n is not the same or a nearest neighbour of m

where $\Phi(x-R_n)$ is a 1s orbital centred on the nth site and as usual * denotes complex conjugate.

Looking at figure 2.9, it appears that a linear combination of atomic orbitals, i.e. of the form

$$\Psi_k(x) = \sum_n C_{nk}\Phi(x - R_n) \qquad 2.20$$

where the C_{kn}, are constants to be determined, might reproduce the wavefunctions of the hypothetical crystal in question, that is the wavefunctions that have a propagating wave character. However, since these must obey Bloch's theorem, the C_{kn} must be of the form

$$C_{kn} = e^{ikR_n} \qquad 2.21$$

To prove this, we substitute 2.21 in 2.20 and observe that the 1-dimensional form of Bloch's theorem is verified. We get, putting $x+R_m$ instead of x in 2.20,

$$\Psi_k(x + R_m) = \sum_n e^{ikR_n} \Phi(x + R_m - R_n) =$$

$$= \sum_n e^{ikR_m} e^{ik(R_n - R_m)} \Phi\big(x - (R_n - R_m)\big) =$$

$$= e^{ikR_m} \sum_n e^{ik(R_n - R_m)} \Phi\big(x - (R_n - R_m)\big) =$$

$$= e^{ikR_m} \Psi_k(x)$$

The wavefunctions as given by 2.20 and 2.21 are not normalized. A normalization constant is required such that

$$\int |\Psi_k|^2 \, dx = \langle \Psi_k, \Psi_k \rangle = 1 \qquad 2.22$$

It is easily verified that the normalization constant required is $1/\sqrt{(N)}$ where N is the number of atoms or unit cells in the chain. Hence

$$\Psi_k(x) = \frac{1}{\sqrt{(N)}} \sum_n e^{ikR_n} \Phi(x - R_n) \qquad 2.23$$

From the Schroedinger equation $H\Psi_k = E_k\Psi_k$, we can obtain the energy E_k by multiplying by Ψ^* and integrating. Then

$$E_k = \int \Psi^* H\Psi_k dx \Rightarrow \qquad 2.24$$

$$\Rightarrow E_k = N^{-1} \sum_n \sum_m e^{ik(R_n - R_m)} \int \Phi^*(x - R_m) H\Phi(x - R_n) dx \qquad 2.25$$

Now in 2.25 we choose to ignore all integrals for which the integers n and m do not denote either the same atom or nearest neighbour atoms. This approximation is essentially the approximation given by equation 2.19. The validity of this approximation derives from the fact that at every point in space either one or both factors of the integrand (the orbitals) are zero. We note that operating by H does not change the extent of the orbital and therefore we can ignore all integrals in which n, m denote atoms that are further apart than the first nearest neighbour distance. Then we can write

$$E_k = \sum_l e^{ikR_l} \int \Phi^*(x - R_l) H\Phi(x)dx \qquad\qquad 2.26$$

where $R_l = R_n - R_m$ denotes first nearest neighbours or zero (if $n=m$). In the linear chain there are only two neighbours of a given atom, one located at a distance a to its right and another at a to its left so $R_l = \pm a$ or 0. Hence

$$E_k = A + \sum_{l=fnn} e^{ikR_l} B = A + 2B\cos(ka) \qquad\qquad 2.27a$$

where fnn means first nearest neighbours,

$$A = \int \Phi^*(x) H\Phi(x)dx \qquad\qquad 2.27b$$

and

$$B = \int \Phi^*(x \pm a) H\Phi(x)dx \qquad\qquad 2.27c$$

This A factor comes from the n=m terms and B from the first nearest neighbours. Note also that by symmetry the + or − terms will give the same B in 2.27c.

There is one more step before we reach our desired conclusion: we write the crystal potential V_{cr} as the sum of the atomic potentials of each atom plus a perturbation that resulted from the bonding of atoms and the redistribution of electron charge that results from it. Then

$$V_{cr} = V_{at} + V_{per} = \sum_i V_{at}(x - R_i) + V_{per} \qquad\qquad 2.28$$

or

$$H = H_{at} + V_{per} \qquad\qquad 2.29$$

Obviously very near the nucleus of the ith atom the total V_{cr} will be equal to $V_{at}(x-R_i)$. Hence

$$A = \int \Phi^*(x)[H_{at} + V_{per}]\Phi(x)dx = E_{at} + C \qquad\qquad 2.30$$

where $E_{at} = \int \Phi^*(x)H_{at}\Phi(x)dx$ = the atomic energy level of the $\Phi(x)$ orbital (i.e. the 1s orbital) in the isolated atom and

$$C = \int \Phi^*(x)V_{per}\Phi(x)dx = \text{a constant}$$

so that equation 2.27a can be written

$$E(k) = E_{at} + C + 2B\cos(ka) \qquad 2.31$$

For each value of k given by 2.12, a corresponding value of energy $E(k)$ will be defined. There will be a maximum value of $E(k)$ equal to $E_{at}+C+2B$ and a minimum equal to $E_{at}+C-2B$. What does equation 2.31 tell us? Exactly what we described in section 2.1 on intuitive grounds: every atomic level E_{at} in an isolated atom splits in a crystal in a band (of width $4B$) centred around a shifted E_{at} (see figures 2.10a and 2.10b). Of course, this simplified model does not hold for most crystals. The nearest solid to which this model can be applied is lithium (Li) which has 3 electrons, 1 electron in the outer 2s orbital and 2 in the inner 1s orbitals, the potential of which can be grouped together with that of its nucleus. Therefore, it looks very much like it is hydrogenic. Like hydrogen, if N is the number of atoms, there will be $2N$ states (counting spin also) and N electrons. The band will be half-filled and therefore Li, having a half filled band, will be a metal, as we will show later in this chapter.

The LCAO method can be extended to take into account more complicated solids that have more than one electron in their outer shell. In particular, we are interested in such semiconductors as Si, Ge, and GaAs, etc., which have two atoms in their unit cell and s and p electrons on each atom. On intuitive grounds we expect each atomic energy level to split into a band. The LCAO method can be an accurate method for these and other materials by including in the summation in equation 2.23 all the orbitals of all the atoms in the unit

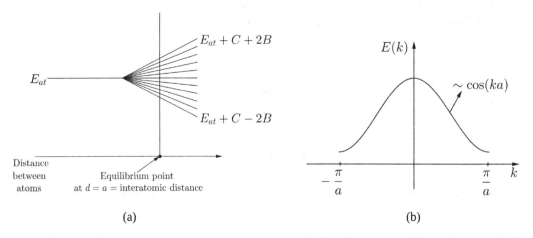

(a)   (b)

FIGURE 2.10 Exact model of the splitting of the 1s energies of the hydrogen chain in a band (a) and their representation in terms of a quasi-continuous wavevector $\boldsymbol{k}$ (b).

cell of the material in question and further interactions than the first nearest neighbours. So, if the positions of the atoms within a unit cell are denoted by the index l, the unit cells by the index n, as previously, and the position vectors of equivalent atoms in different unit cells by $\mathbf{R}_{n,l}$, equation 2.23 can be generalized in 3 dimensions for more than 1 atom in the unit cell and more than one kind of orbital as follows:

$$\Psi_k(\mathbf{r}) = \sum_{j,l} a_{j,l} \sum_n e^{i\mathbf{k}\cdot\mathbf{R}_{n,l}} \Phi_j(\mathbf{r} - \mathbf{R}_{n,l}) = \tag{2.32}$$

$$= \sum_{j,l} a_{j,l} \tilde{\Psi}_{j,l}(\mathbf{k}) \tag{2.33}$$

where $a_{j,l}$ are the expansion (or weighting) coefficients of the $\Phi_{j,l}$ orbital which is of type j (i.e. s or p) and is centred on the equivalent atoms at positions $\mathbf{R}_{n,l}$.

Equation 2.32 does not lead to a simple expression of the form $E(k)$= a known quantity. Instead it leads to a set of linear equations. To see this, we substitute the expansion of 2.33 into the Schroedinger equation $H\Psi=E\Psi$ to obtain

$$\sum_{j,l} a_{j,l} H \tilde{\Psi}_{j,l}(\mathbf{k}) = E \sum_{j,l} a_{j,l} \tilde{\Psi}_{j,l}(\mathbf{k})$$

Now following the standard procedure of linear algebra, we multiply on the left by $\tilde{\Psi}^*_{j'l'}(\mathbf{k})$ and integrate. We get, hiding the $\mathbf{k}$ dependence for the moment,

$$\sum_{j,l} a_{j,l} \int \tilde{\Psi}^*_{j',l'} H \tilde{\Psi}_{j,l} dV = E a_{jl} \int \tilde{\Psi}^*_{j',l'} \tilde{\Psi}_{j,l} dV \tag{2.34}$$

$$\Rightarrow \sum_{j,l} a_{j,l} H_{jl,jl'} = a_{jl} E S_{jl,jl'} \Rightarrow$$

$$\Rightarrow \sum_{j,l} a_{jl} \left(H_{jl,jl'} - E S_{jl,jl'} \right) = 0 \tag{2.35}$$

where

$$H_{jl,jl'} = \int \tilde{\Psi}^*_{j'l'} H \tilde{\Psi}_{j,l} dV$$

and

$$S_{jl,jl'} = \int \tilde{\Psi}^*_{j'l'} \tilde{\Psi}_{j,l} dV$$

and V is the volume of the crystal.

For the set of linear equations 2.34 in the unknowns a_{jl} to have a non-trivial (i.e. non-zero) solution we must have

$$det\left(H(\boldsymbol{k})-E(\boldsymbol{k})S\right)=0 \qquad\qquad 2.36$$

where we have reinstated the $\boldsymbol{k}$ dependence and have treated $H_{jl,j'l'}$ and $S_{jl,j'l'}$ as the matrix elements of the corresponding matrices H and S. Equation 2.36 is called the secular equation and the reader would recognize the above equations as a standard procedure for turning a differential equation into an algebraic one. The above equations would only be a mathematical trick with no computational significance if the summation in 2.34 were not over but a small number of values of j and l. Indeed, for most semiconductors (not all) we only need to consider the s and the three p type of orbitals with only two atoms in the unit cell so that the size of the matrices H and S is manageable, although, in principle, one would have to include in 2.32 an infinite sum of orbital types.

The bands that we get from the secular equation are radically different from the simple model of the linear chain with only one s orbital in the unit cell. The main feature of the realistic model that led to the secular equation is that we do not get only one $E(\boldsymbol{k})$ curve as in figure 2.10 but many $E(\boldsymbol{k})$ curves as the size of the matrices in 2.36 since the determinant of a matrix gives an algebraic equation of order equal to the size of its matrix. So we should write $E_m(\boldsymbol{k})$ where m is a band index referring to the (j,l)th eigenvalue of the secular equation. The same holds for the wavefunctions. We should write $\Psi_{mk}(\boldsymbol{r})$ and not $\Psi_k(\boldsymbol{r})$. Figures 2.11a–c shows the bands in Al, Si, and GaAs respectively. Given that $\boldsymbol{k}$ is a

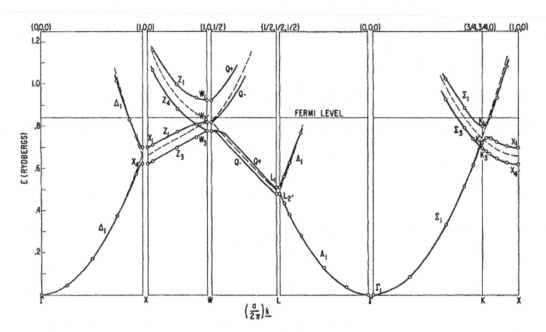

FIGURE 2.11a The band structure of aluminum, calculations by B. Segall, Phys. Rev. 124, 1797 (1961). Capital letters (English or Greek) denote points in the Brillouin Zone.

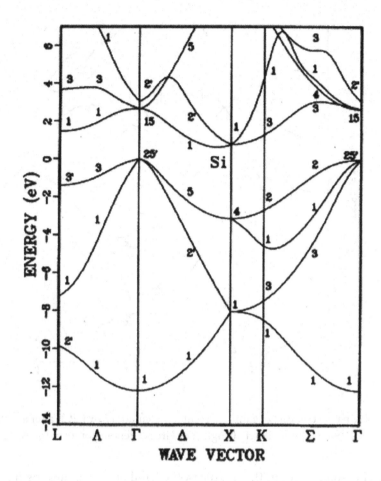

FIGURE 2.11b   The band structure of silicon, calculations by C. S. Wang and B. M. Klein, Phys. Rev. B 24, 3393 (1981). Capital letters (English or Greek) denote points in the Brillouin Zone.

3-dimensional vector, the 1-dimensional vector k appearing in figure 2.11 portrays the values of k along the axes of high symmetry of the Brillouin zone (BZ) shown in figure 2.8—all 3 materials have the same BZ. So far we have not linked the theory with the properties of the solid in question. We have merely stated without a proper explanation that partially filled bands give metallic behaviour. We are now ready to tackle the questions we posed ourselves at the beginning of this chapter (see section 2.1).

The most trivial question we have so far asked is why Al is a metal and GaAs is a semiconductor. It is a property of a partly filled band that it can exhibit conductivity, whereas a completely filled band gives a zero conductivity. Why is this so? As an electric field is applied to a solid, the electrons can be accelerated and give rise to electric current, but only if they can accept the energy of the electric field. To do so they must be in states which have empty states just above them (in energy) to which they can jump as they are being accelerated. Electrons in completely filled bands can't absorb energy from the electric field because there are no such empty states available. Solids in which the highest occupied band is partially filled fulfill the above condition and are metals. We note that in metals

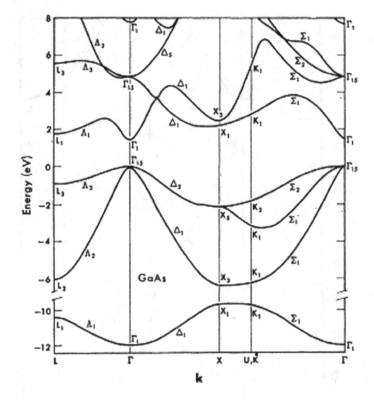

FIGURE 2.11c   The band structure of gallium arsenide, calculations by J. P. Walter and M. L. Cohen, Phys. Rev. 183, 763 (1969). Capital letters (English or Greek) denote points in the Brillouin Zone.

the Fermi level is somewhere in the middle of the band and separates the filled from the empty states at $T = 0$.

Semiconductors are solids in which the highest occupied band, called the valence band, is completely filled at $T = 0$, but it is separated from the next empty band by a band gap $E_g < 2eV$. Again at $T = 0$ these materials can't exhibit conductivity but as the temperature T rises, electrons can be excited thermally from the valence band to the next empty band called conduction band, thus making both bands conductive. This is the case of GaAs, for example, in which the minimum of the conduction band occurring at the Γ point of the BZ is separated by a gap of $1.4eV$ from the maximum of the valence band occurring also at Γ. But how do we know that the highest occupied band at T=0 (i.e. the valence band) is full and the one above it (i.e. the conduction band) is completely empty? This is a mandatory requirement because in either case we will not have a semiconductor but a metal. The answer is that we count the number of states and the number of electrons and then we decide. The procedure is easy and requires only simple arithmetic. We describe this for GaAs immediately below. The arguments are valid only at T=0, but this is the only temperature we need to examine to deduce the character of the material.

Let us draw a figure similar to 2.10a specifically for GaAs. The latter has one Ga and one As atom in its unit cell. We only need to consider the outer shell of each atom. The orbitals of the inner shells of any atom give completely filled bands. The electronic structure of the

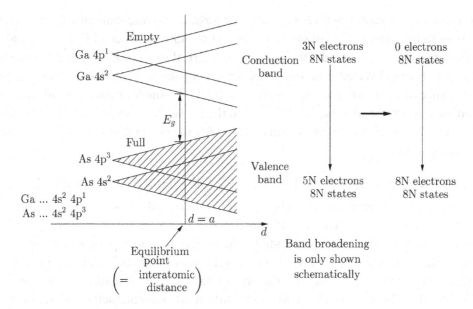

FIGURE 2.12 Schematic formation and filling of bands in GaAs as deduced from the atomic structure of the constituent atoms.

outer shell of Ga is $4s^2 4p^1$, meaning two electrons in its 4s orbital and one in its 4p orbital, while that of As is $4s^2 4p^3$. The broadening of these atomic levels into bands is shown schematically in figure 2.12. Now according to Bloch's theorem, each such atomic level gives a band with N such levels in the solid where N is the number of unit cells. We must not forget that each such atomic level has a degeneracy of $2(2l+1)$ in the atom itself where l is the second quantum number of the specific orbital. We remind the reader (see also equation 1.39) that s-type orbital means l=0 and p-type orbital means l=1. We can then construct the following table of bands with electrons and available states in each band.

We can see from figure 2.12 that the valence band of GaAs arises from the splitting of the As atomic levels and the conduction band from the Ga atomic levels. From the table above (table 2.1), it appears as if the valence band contains $5N$ electrons and the conduction band $3N$ electrons, but this is misleading because what actually will happen is that the $3N$ electrons of the Ga states will fall into the empty $3N$ states of the valence band, filling it completely so that the valence band is full and the conduction band is empty. This type of solid is either a semiconductor or an insulator depending on the value of the

TABLE 2.1   Number of states and electrons in the bands of GaAs

Atomic Level	Number of Electrons		Number of States	
Ga 4s l=0	$2N$	$\left.\right\}3N$	$8N\left\{\right.$	$2N$
Ga 4p l=1	N			$6N$
As 4s l=0	$2N$	$\left.\right\}5N$	$8N\left\{\right.$	$2N$
As 4p l=1	$3N$			$6N$

energy band gap E_g. Solids with $E_g < 2eV$ are categorized as semiconductors. Higher values of E_g refer to insulators. We will see in a later section how the value of E_g is related to the number of electrons in a band. Of course the numerical value of E_g or even the existence of a gap cannot be deduced from the simple argument above. Note that we have assumed the existence of a gap in figure 2.12 with no validation—the four bands could have converged into a single band—that is why we need the LCAO (and other methods) of calculation. There is a lot of important information hidden in the $E_m(k)$ curves as we will see in a following section.

2.4 QUICK REVISION OF THE CONCEPT OF A HOLE AND DOPING

The following constitutes a quick revision of early undergraduate knowledge on semiconductors. This book assumes that a basic knowledge of electrons, holes, and doping has been obtained during the first or second undergraduate years. But for the sake of completeness of this book, it is very briefly repeated here. This section will also help clarify some misconceptions present in the early stages of student familiarization with semiconductor concepts.

We will use silicon (Si)—the most commonly used semiconductor—as an example. Silicon, in the form of an isolated atom, has the electronic structure $3s^2 3p^2$, i.e. it has 4 electrons in its outer shell and 2 atoms per unit cell. Following the methodology of the previous section, we can easily deduce that its valence band is completely full and its conduction band is completely empty at $T=0$. This is schematically shown in figure 2.13a. A 2-dimensional simplified version of the bonding is shown in figure 2.13b. The electrons in the valence band in 2.13a are the ones that participate in the bonds shown in figure 2.13b.

As the temperature increases, electrons from the valence band are excited (jump) into the conduction band, the latter filling with electrons, while in the valence band, states are being emptied. Everything is being done according to the Fermi–Dirac statistics. The conductivity at $T=0$ was zero but at a finite or room temperature T, it acquires a small value.

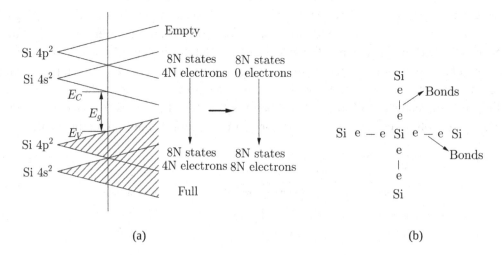

FIGURE 2.13 (a) Complete filling of the valence band of Si by the electrons of the $Si-Si$ bonds (b) A 2-dimensional schematic view of bonds in Si.

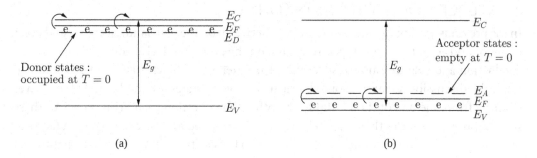

FIGURE 2.14   Donor states (a) and acceptor states (b) in Si and in semiconductors in general.

The increase in conductivity does not only come from the conduction band, which now contains some electrons, but also from the empty states in the valence band which are called holes. Note that in an early stage of teaching these are portrayed as missing electrons in bonds and as actual holes in bonds. However, these states are empty Bloch states which extend from the one end of the crystal to the other end and hence can propagate through the crystal. It is slightly misleading to think of them as actual holes jumping from one bond to another.

Now if an As atom, which has 5 electrons in its outer shell, is substituted for a Si atom, all 4 bonds will be satisfied and there will be an extra electron present. This extra electron occupies at T=0 an electronic state of energy E_D which lies just a few tens of a meV below E_C as shown in figure 2.14a. Its wavefunction is not a Bloch state (since it lies in the band gap of the semiconductor) but it is localized around the As atom resembling (but not exactly) an As orbital hence the electron in it can not conduct. Since its energy level is so close to the conduction band E_C, this electron can be excited to the conduction band by thermal excitation and then participate in the conduction process.   At room temperature all As atoms, called donors, are ionized. Note that KT=26meV at room temperature where K is Boltzmann's constant.

If, on the other hand, we substitute a P (phosphorus) atom for a Si atom, which has 3 electrons in its outer shell, then there will be a missing electron to satisfy the fourth bond. This will create an actual hole, i.e. a localized state resembling an atomic P orbital whose energy level E_A lies close to the valence band edge as shown in figure 2.14b. The state is obviously empty at T=0, but again given that E_A_E_V is of the order of tens of meV, it can be filled by an electron of the valence band by thermal excitation at T>0, leaving a hole in the valence band. The P atoms are called acceptors and their localized states are called acceptor states. We wish to make clear the sharp distinction of the holes in the valence band with the acceptor states. The latter are localized near the P atoms and are empty at T=0. At T>0 they are filled with electrons, but these electrons can't participate in the conduction process. On the contrary, the holes in the valence band being in Bloch states can move freely and increase the conductivity of the solid just like the electrons in the conduction band which are likewise in Bloch states. For the calculation of these donor and acceptor states, see Appendix C.

2.5 VELOCITY OF ELECTRONS IN SOLIDS

In the previous section we made a distinction between, on the one hand, electrons and holes in donor and acceptor states respectively and, on the other hand, electrons and holes in the conduction and valence bands respectively. The latter are in Bloch states and can conduct. From the discussion so far, it should be clear that Bloch states are modulated plane waves running from one end of the crystal to the other. The question then arises how can these states represent particles that are accelerated, then scattered and accelerated again (according to a simple picture of conduction), i.e. how can the electrons have a localized nature but extend all over the crystal. We will delve into this question in chapter 4, but for the moment it suffices to say the following. When an electric field is applied to a solid, the wavefunctions of the electrons are no longer Bloch states, $\Psi_{m,k}$, but are wavepackets made out of Bloch states which are localized in space. Each wavepacket is sharply peaked around a particular Bloch state with wavevector k. Mathematically, these wavepackets can be written $\Psi(r)$ as

$$\Psi(r,t) = \int_{BZ} A(k)\Psi_{m,k}(r)e^{iEt/h}d^3k \qquad 2.37$$

where the $A(k)$ are peaked around a given k as shown in figure 2.15a. BZ stands for the Brillouin zone. The space variation of $\Psi(r)$ is shown in figure 2.15b. Both figures give a 1-dimensional representation of these variables at an instant of time.

Now the motion of a wavepacket is determined by its group velocity which from the theory of simple optics is

$$v_g = \frac{d\omega}{dk} \text{ (in one dimension)} \qquad 2.38$$

But in quantum mechanics, we have the relationship

$$E = \hbar\omega$$

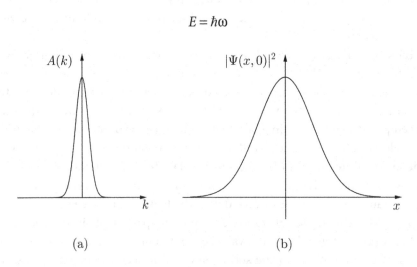

(a)   (b)

FIGURE 2.15 Weighting factors of the Bloch waves in a wavepacket $\Psi(x,t=0)$ (a) and $|\Psi(x,0)|^2$ itself (b).

Hence the motion of a wavepacket, each composed of wavefunctions $\Psi_{m,k}$ with an energy $E_m(k)$, where m is the band index, is given in one dimension by the group velocity

$$v_g^m(k) = \frac{1}{\hbar} \frac{dE_m(k)}{dk} \qquad \text{2.39a}$$

and in 3 dimensions is given by

$$v_g^m(k) = \frac{1}{\hbar} \nabla_k E_m(k) \qquad \text{2.39b}$$

where ∇_k is the grad in k space.

We can immediately verify an important conclusion we drew previously. A completely filled band cannot exhibit a current. The current density J is given by

$$J = \sum_i e v_i = \sum_k \frac{e}{\hbar} \nabla_k E(k) \qquad \text{2.40}$$

where the summation over k is over the allowed k values in the Brillouin zone. However an examination of figure 2.8 shows that the BZ of the FCC lattice is symmetric around zero, i.e. for every allowed k there is always an allowed $-k$, with $\nabla_{-k} E(-k) = -\nabla_k E(k)$. This is a property of all BZs. We have already encountered this property in the $E(k)$ expression for the simple linear chain of the hydrogenic atoms, equation 2.31. Therefore, the sum in equation 2.40 above is zero. It is worth noting that while donor atoms fill an empty conduction band with electrons, thereby increasing its conductivity, the acceptors do exactly the same thing, i.e. increase the conductivity by partially emptying the valence band, however odd it may seem at first glance.

Another point we would like to discuss is the notion of "crystal momentum". The quantity $\hbar k$ plays the role of momentum as far as the external forces are concerned, i.e. we can write

$$F = e\mathcal{E} = \hbar \frac{dk}{dt} \qquad \text{2.41}$$

There is a formal proof of equation 2.41, which we wil not give, but the reader may be satisfied by noting that, on intuitive grounds, Fdt is the impulse in time dt and hence this quantity must give the corresponding change in momentum $d(\hbar k)$. We note however that in 2.41 the electrons must be considered as wavepackets as already described and the k refers to the wavevector of the dominant Bloch component. It must be stressed though, that the Bloch functions are not eigenfunctions of the momentum operator and hence $\hbar k$ would not be a result of a momentum measurement in a semiconductor. Equation 2.41 merely relates the **external** force applied to a semiconductor to the rate of change of the **k** vector of an electron.

2.6 THE CONCEPT OF EFFECTIVE MASS

An examination of the $E(\mathbf{k})$ curves of GaAs, figure 2.11, reveals that both the conduction band near its bottom E_C, and the valence band near its top E_V, occurring both at $k=0$, are very nearly parabolic and isotropic. Therefore, their energies can be represented by the modulus k of $\mathbf{k}$ only. The energy regions around E_C and E_V are very important because these are the energies that electrons from donors and holes from acceptors will respectively occupy. But these energy ranges are only a small fraction of the whole band so it is acceptable that we should find a simpler representation than the numerical output of a rigorous calculation. This can be done by using Taylor's theorem. Expanding therefore up to 2nd order around E_C for the conduction band and around E_V for the valence band of GaAs we have

$$E_{con}(k) = E_C + \frac{1}{2}\frac{d^2 E_{con}(k)}{dk^2}\bigg|_{k=0} k^2 \qquad 2.42$$

and

$$E_{val}(k) = E_V + \frac{1}{2}\frac{d^2 E_{val}(k)}{dk^2}\bigg|_{k=0} k^2 \qquad 2.43$$

Note that in 2.42 and 2.43, as already noted, we have assumed that both bands depend only on the magnitude of $\mathbf{k}$. Also note that there are no first order terms in 2.42 and 2.43 because the expansions have been performed around a minimum for the conduction band and a maximum for the valence band respectively.

The above relations remind us of the free electron case or the particle in a box problem where in both cases the energy is of the form $E = \hbar k^2/2m$. Now for parabolic isotropic bands the derivatives appearing in 2.42 and 2.43 are constants. It is appealing therefore to rewrite 2.42 and 2.43 in a way that is reminiscent of the above energy–wavevector relation, i.e. put 2.42 and 2.43 in the form

$$E_{con}(k) = E_C + \frac{\hbar^2 k^2}{2m_n^*} \qquad 2.44$$

$$E_{val}(k) = E_V + \frac{\hbar^2 k^2}{2m_p^*} \qquad 2.45$$

where

$$m_n^* = \hbar^2 \left[\left(\frac{d^2 E_{con}(k)}{dk^2}\right)\bigg|_{k=0}\right]^{-1} \qquad 2.44a$$

and

$$m_p^* = \hbar^2 \left[\left(\frac{d^2 E_{val}(k)}{dk^2} \right) \Bigg|_{k=0} \right]^{-1}$$ 2.45a

The parameters m_m^*, m_p^* are called the effective mass of electrons and holes respectively and indeed have units of mass so that 2.44 and 2.45 can be reinterpreted as total energies composed of a constant potential energy (E_C in 2.44 and E_V in 2.45) plus a kinetic energy term. Note that the effective mass of the valence band, m_p^*, is negative (all states lie below E_V) so that it is customary, in order to retain the concept of mass, to write

$$E_{val}(k) = E_V - \frac{\hbar^2 k^2}{2m_p^*}$$ 2.45b

where $m_p^* > 0$.

An examination of the Si energy bands in figure 2.11 reveals that the conduction band minimum E_C and the valence band maximum E_V do not occur at the same k point. In particular, E_V occurs at the centre of the BZ, point labelled Γ, whereas the conduction minimum occurs on the ΓX axis (see figure 2.8). Furthermore, the conduction band is not isotropic near E_C, see figure 2.11, so that a slightly more complicated expression is needed than that of equation 2.44 for Si.

$$E_{con}(k_t, k_l) = E_C + \frac{\hbar^2 k_t^2}{2m_t^*} + \frac{\hbar^2 k_t^2}{2m_t^*} + \frac{\hbar^2 k_l^2}{2m_l^*}$$ 2.46

where k_l is the (longitudinal) component of k along the ΓX axis and k_t the other two transverse to this components, measured from the point of expansion. That is, the constant energy surfaces, in k space, of the conduction band in Si are elongated ellipsoids, as shown in figure 2.16. There are 6 of these ellipsoids corresponding to the positive and negative portions of the 3 k axes.

In equation 2.46 we have treated m_t^* and m_l^* as simple parameters with no formal definition as in equations 2.44a and 2.45a. The effective mass can be generalized to fit any material by making it a tensor as follows.

$$m_{ij}^* = \hbar^2 \left(\frac{\partial^2 E}{\partial k_i \partial k_j} \right)^{-1}$$ 2.47

Then all of the known semiconductor band structures can be expanded near the minimum E_C of the conduction band as follows

$$E_{con}(k) = E_C + \frac{\hbar^2}{2} \left(\frac{k_x^2}{m_x^*} + \frac{k_y^2}{m_y^*} + \frac{k_z^2}{m_z^*} \right)$$ 2.48

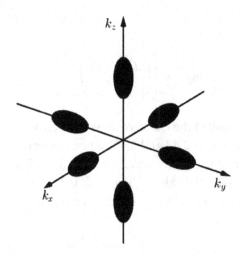

FIGURE 2.16   The regions in **k** space where the conduction band minima of Si are located: they are ellipsoidal in shape.

where k_x, k_y, k_z are the components of **k** measured from the point of expansion and m_x, m_y, m_z are the diagonal elements of the effective mass tensor. The valence band maximum on the other hand is usually isotropic.

The concept of effective mass is not only a device which allows simple expansions of the bands near their minimum or maximum. It also allows a description of electrons in a more classical manner as can be seen by the following simple 1-dimensional proof. Imagine an electron in a conduction band with an isotropic effective mass m^* moving under the action of an electric field $\mathcal{E}$. The electron will move in the opposite direction from which the electric field is applied. So to avoid minus signs we measure the distance travelled ds in time dt in this direction. Then the energy gained under the action of the field is (with v being the velocity)

$$dE = e\mathcal{E}ds = e\mathcal{E}v(k)dt \Rightarrow$$

$$\Rightarrow dE = e\mathcal{E}\frac{1}{\hbar}\frac{dE}{dk}dt \Rightarrow e\mathcal{E} = \hbar\frac{dk}{dt}$$

This is actually a 1-dimensional version of equation 2.41, so it would have been proper to start from there, but we warn the reader that the above mathematical steps do not constitute a proof of equation 2.41. However, from this point onwards there are no loopholes in our derivation. We get

$$e\mathcal{E} = \hbar\frac{dk}{dt} = \hbar\frac{dk}{dv}\frac{dv}{dt} = \hbar\left(\frac{dv}{dk}\right)^{-1}\frac{dv}{dt} \Rightarrow$$

$$e\mathcal{E} = \hbar^2\left(\frac{d^2E}{dk^2}\right)^{-1}\frac{dv}{dt} \Rightarrow e\mathcal{E} = m^*\frac{dv}{dt} \qquad \text{2.49a}$$

Equation 2.49 looks like Newton's third law where the force is proportional to the acceleration. Indeed, electrons in the conduction band (or holes in the valence band of a semiconductor) can be thought of as classical particles moving under the action of external forces with the effective mass of the band, however, substituted for the real mass. The 3-dimensional version of equation 2.49a is

$$\frac{dv_i}{dt} = \frac{1}{\hbar^2}\frac{\partial^2 E(\boldsymbol{k})}{\partial k_i\,\partial k_i} e\mathcal{E}$$

2.49b

We have now reached a stage where we can draw a very important conclusion relating band theory to applications: the lighter the effective mass of the electron in a semiconductor compared to another, the faster these electrons will move and the faster will they respond to signals. As an example, note that the effective mass of electrons in GaAs is $0.067m$ whereas in Si it is $0.92m$ where m is the vacuum mass of electrons. So substantial differences occur between semiconductors which may also explain the interest in GaAs in particular.

2.7 CONCENTRATION OF CARRIERS IN SEMICONDUCTORS AND METALS

We are now ready to evaluate the densities of carriers in the bands of semiconductors which form one of the main factors of their classical conductivity. Let us consider first the number of carriers per unit volume in the conduction band of a semiconductor. If P_{FD}, the Fermi–Dirac probability of occupation of a state in a conduction band is, say, 0.7, that means that on average there are 0.7 electrons in that state, so it is correct to write that the number of electrons per unit volume, n, in that band is

$$n = \frac{1}{V}\sum_{E_k > E_C} P_{FD}(E_k)$$

2.50

where V is the volume of the semiconductor. The problem with equation 2.50 is that while E_k is a discrete quantity, there are roughly 10^{23} energies per mole to be counted so that it would be beneficial to convert the sum in 2.50 into an integral. To achieve that we only need to count the number of energy levels $\Delta N(E)$ between E and $E+\Delta E$. Then the sum can be converted into an integral as follows

$$n = \frac{1}{V}\sum_{\Delta E}\Delta N(E)P_{FD}(E) =$$

$$= \int_{E_C}^{\infty} g_n(E)P_{FD}(E)dE$$

2.51

where

$$g_n(E)\Delta E = \frac{\Delta N(E)}{V}$$

2.51a

is the number of states per unit volume between E and $E+\Delta E$ in the conduction band or alternatively $g_n(E)$ is the density of states per unit energy per unit volume in the conduction band.

In the limit of $\Delta E \to 0$, $g(E)$ can be thought as the derivative of $N(E)$, where the latter is defined as the number of states up to the energy E. Although the energies E_k are discrete, as the subscript k signifies, they are so dense that they can be treated as a continuum. The task is then to find an expression for $g(E)$. We have seen in the previous paragraph that most semiconductors of interest can have their conduction band expanded about its minimum as (cf. equation 2.48 which we reproduce here)

$$E - E_C = \frac{\hbar^2 k_x^2}{2m_x^*} + \frac{\hbar^2 k_y^2}{2m_y^*} + \frac{\hbar^2 k_z^2}{2m_z^*} \qquad 2.48$$

that is, the constant energy surfaces in k space are in general ellipsoids. We remind the reader that in Si the conduction band minima E_C occur (see figure 2.16) along the ΓX axis. Two effective masses are equal in 2.48 and are labelled m_t, and the third one is labelled m_l, see equation 2.46. In GaAs with E_C occurring at $k=0$ all effective masses are equal. However, for the sake of generality we will proceed with the calculation of $g(E)$ assuming all m^* are different.

The volume of an ellipsoid of the form

$$\frac{x^2}{a^2} + \frac{y^2}{b^2} + \frac{z^2}{c^2} = 1$$

is $V = \frac{4}{3}\pi abc$. So the volume in k space of an ellipsoid of the form of equation 2.48, enclosed by a surface of energy E is

$$V(E) = \frac{4}{3}\frac{\pi\sqrt{8m_x^* m_y^* m_z^*}}{\hbar^3}(E - E_C)^{3/2}$$

Therefore, an incremental volume dV enclosed between ellipsoids of energy E and $E+dE$ is (taking the derivative)

$$dV(E) = 4\frac{\pi\sqrt{2m_x^* m_y^* m_z^*}}{\hbar^3}(E - E_C)^{1/2}\,dE$$

But we have seen in section 2.2 (see equation 2.12) that in 1 dimension, the distance between k points is $\frac{2\pi}{L}$, and in 3 dimensions, it is $\left(\frac{2\pi}{L}\right)^3$ for a cubic sample of size L. Hence the number of states $\Delta N(E)$ between E and $E+dE$ will be given by the elemental volume $dV(E)$ divided by the above spacing

$$\Delta N(E) = \frac{L^3 4\pi \sqrt{2m_x^* m_y^* m_z^*} (E - E_C)^{1/2} dE}{(2\pi)^3 \hbar^3}$$

so that

$$g_n(E)dE = \frac{4\pi \sqrt{2m_x^* m_y^* m_z^*}}{h^3} (E - E_C)^{1/2} dE \qquad 2.52$$

where in 2.52 we have divided by the volume $V=L^3$ and have simplified the denominator, noting that $(2\pi)^3 \hbar^3 = h^3$. Equation 2.52 is not the final result for the density of states (usually abbreviated DOS). There are two additional multiplicative factors: first as we have noted earlier, there are six such ellipsoids in Si so we have to multiply by this number. We assign the symbol f to refer to the number of minima in any semiconductor. Second, we have to take care of spin, every Bloch state can carry two electrons. Hence the final result is

$$g_n(E)dE = \frac{8\pi f \sqrt{2m_x^* m_y^* m_z^*}}{h^3} (E - E_C)^{1/2} dE \qquad 2.53$$

Therefore, from 2.51 we get the concentration of electrons in the conduction band

$$n = \int_{E_C}^{\infty} \frac{8\pi f}{h^3} \frac{\sqrt{2}(m_n^*)^{3/2} (E - E_C)^{1/2}}{1 + exp\left(\dfrac{E - E_F}{KT}\right)} dE \qquad 2.54$$

where $m_n^* = (m_x^* m_y^* m_z^*)^{1/3}$. We remind the reader that K is Boltzmann's constant. If the 1 in the denominator of the above integral can be ignored, then the integral can be calculated analytically and the result is

$$n = N_C exp\left(\frac{E_F - E_C}{KT}\right) \qquad 2.55$$

where $N_C = 2f\left(\dfrac{2\pi m_n^* KT}{h^2}\right)^{3/2}$ is called the effective density of states. If, on the other hand, the Fermi level is very close to the conduction band edge E_C, the 1 in the denominator can't be ignored and the integral has to be evaluated numerically. Equation 2.54 can then be rewritten as

$$n = N_C F_{1/2}\left[\frac{(E_C - E_F)}{KT}\right] \tag{2.56}$$

where the so-called Fermi integral $F_{1/2}$ of order ½ is defined as

$$F_{1/2}\left(\frac{E_C - E_F}{KT}\right) = \frac{2}{\sqrt{\pi}}\int_0^\infty \frac{\left(\dfrac{E - E_C}{KT}\right)^{1/2} d(E/kT)}{1 + exp\left(\dfrac{E - E_F}{KT}\right)} \tag{2.56a}$$

The above formulae 2.56 and 2.56a are necessary only when the Fermi energy E_F moves inside the conduction band.

A completely parallel argument shows that the concentration of holes p in the valence band

$$p = \int_{-\infty}^{E_V} g_p(E)(1 - P_{FD}(E))dE \tag{2.57}$$

where $g_p(E)$, the density of states in the valence band, is (again provided that E_F is not near E_V)

$$p = N_V exp\left(\frac{E_V - E_F}{KT}\right) \tag{2.58}$$

$$\text{where } N_V = 2\left(\frac{2\pi m_p^* KT}{h^2}\right)^{3/2} \tag{2.58a}$$

The degeneracy of the band edge is taken as $f=1$ in this case because E_V is situated at $k=0$ for Si and most other semiconductors.

Consider now an intrinsic semiconductor (i.e. no doping present). For every electron that is excited into the conduction band, a hole is left in the valence band. Hence $n=p$. Substituting into $n=p$ the expression for n from equation 2.55 and the expression for p from equation 2.58 and solving for E_F, we get

$$E_F = \frac{E_C + E_V}{2} - \frac{KT}{2}\ln\left(\frac{N_C}{N_V}\right) \tag{2.59}$$

The second term in 2.59 is negligible compared to the first, (remember at room temperature $KT=26meV$). So to a very good approximation $E_{F_i}=$the Fermi level of the intrinsic semiconductor is $E_{F_i} \simeq \dfrac{(E_C + E_V)}{2}$. We note that $E_C - E_{F_i} > KT$ and therefore the approximate expressions 2.55 and 2.58 were correctly used to obtain E_{F_i}.

By multiplying 2.55 and 2.58 we get for an intrinsic semiconductor

$$n_i = p_i = \sqrt{N_C N_V}\, exp\left(-E_g/2KT\right) \qquad 2.60$$

where E_g is the band-gap $E_C - E_V$ or

$$n \cdot p = n_i^2 \qquad 2.61$$

Since the product np is independent of the Fermi level, which (as we shall see) is sensitive to the doping of a semiconductor, the above equation holds for any level of doping of a semiconductor. Equation 2.61 is usually called the law of mass action. Equations 2.55 and 2.58 can now be rewritten in terms of E_{F_i} and n_i

$$n = n_i exp\left(\frac{E_F - E_{F_i}}{KT}\right) \qquad 2.62$$

and

$$p = n_i exp\left(\frac{E_{F_i} - E_F}{KT}\right) \qquad 2.63$$

Now consider an N-type semiconductor. As explained in section 2.4, for every donor atom in the crystal a new energy level E_D is created just below E_C (see figure 2.14), which is occupied by the extra electron of the donor atom at $T=0$ and which is progressively emptied (the donor atom is being ionized) as the temperature T increases. The electrons at E_D jump into the conduction band. What can we say about the Fermi level E_F of such a system if the states at E_D are occupied at $T=0$ and the ones in the conduction band are empty? The Fermi level at $T=0$ must lie in between E_D and E_C since E_F separates the full from the empty states (at $T=0$). In fact it lies midway at $(E_D + E_C)/2$. As the temperature increases E_F moves down the gap since the levels E_D are emptied. An exactly parallel argument holds for a P-type semiconductor. The Fermi level E_F lies below E_A at $(E_V + E_A)/2$ and as the levels at E_A are being filled and holes are created in the valence band, E_F moves up the gap. In both cases of donor and acceptor doping if the temperature is raised at such levels that the number of electrons and holes created by thermal excitation from the valence to the conduction band exceeds the number created by doping, the Fermi level returns to the mid-point of the energy gap. The movement of E_F with temperature in both N- and P-type semiconductors is shown in figures 2.17a,b.

Since any semiconductor N- or P-type must remain neutral we must have

$$n + N_A^- = p + N_D^+ \qquad 2.64$$

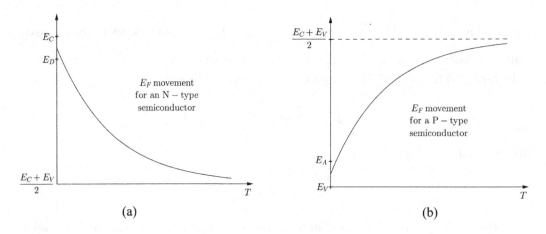

FIGURE 2.17 Movement of the Fermi level E_F as the temperature increases in an N-type (a) and P-type (b) semiconductor.

The Fermi level E_F of any kind of semiconductor can be calculated as follows: divide equation 2.62 by 2.63, we then get

$$\frac{n}{p} = exp\left(\frac{2E_F - 2E_{F_i}}{KT}\right) \Rightarrow$$

$$\Rightarrow \frac{2}{KT}\left(E_F - E_{F_i}\right) = ln\left(\frac{n}{p}\right) \Rightarrow$$

$$\Rightarrow E_F = E_{F_i} + \frac{KT}{2}ln\left(\frac{n}{p}\right) \qquad 2.65$$

When we have an N-type semiconductor, $n \gg p$ and its Fermi level E_{F_n} lies in the upper half of the band-gap and conversely for a P-type semiconductor $p \gg n$ and its Fermi level E_{F_p} lies in the bottom half of the band-gap. The ionization process of both types of impurities is complete at room temperature because the impurity levels E_D or E_A happen to be very close to the respective band-edge E_C or E_V. Therefore, at room temperature

$$N_D^+ = N_D \qquad 2.66a$$

$$\text{and } N_A^- = N_A \qquad 2.66b$$

The doping of most semiconductors is such that $n_i \ll N_D$ or $n_i \ll N_A$. Using the relation 2.61 we also have for an N- and P-type semiconductor respectively (i.e. for the minority carriers)

$$p = n_i^2/N_D \qquad 2.67a$$

$$\text{and } n = n_i^2/N_A \qquad 2.67b$$

Hence we get by virtue of 2.64 that at room temperature for an N-type semiconductor to a very good approximation

$$n \approx N_D \qquad\qquad 2.68a$$

and for a P-type semiconductor

$$p \approx N_A \qquad\qquad 2.68b$$

Substitution of the above values for n and p in the general relation 2.65 for the Fermi levels of an N-doped and a P-doped semiconductor respectively leads to

$$E_{F_n} = E_C + KTln\left(\frac{N_D}{n_i}\right) \qquad\qquad 2.69a$$

$$E_{F_p} = E_C - KTln\left(\frac{N_A}{n_i}\right) \qquad\qquad 2.69b$$

Now that we have dealt with semiconductors, the evaluation of the corresponding quantities in metals is straightforward. The number of electrons per unit volume n in the conduction band of a metal can be evaluated by integrating the density of states (DOS) multiplied by the Fermi–Dirac probability from the bottom of the band to infinity. The DOS is the same for metals as for semiconductors to the extent that both can be treated as parabolic (i.e. $\propto k^2$). Furthermore, we can choose as our zero of energy the bottom of the conduction band so that

$$g_m(E) = \frac{4\pi\sqrt{m^*}}{h^3} E^{1/2} \qquad\qquad 2.70$$

where $g_m(E)$ is the DOS of metals at E. Then

$$n = \int_0^\infty \frac{g_m(E)dE}{1 + exp\left(\dfrac{E - E_F}{KT}\right)} \qquad\qquad 2.71$$

The above integral is easily evaluated at $T = 0$ when the denominator of the integrand is 1 and the upper limit of the integral will be E_F. We get

$$n = \frac{(2mE_F)^{3/2}}{3\pi^2\hbar^3} \qquad\qquad 2.72$$

The above formula may also be used for $T =$ room temperature because the Fermi level of a metal does not move significantly compared to the width of the conduction band of a metal. We remind the reader that $KT \approx 26meV$ and the width of the bands in metals are of the order of $10eV$.

2.8 THE EFFECTIVE MASS EQUATION

The concept of effective mass is not only useful for classical concepts like velocity and acceleration, but it can also be used to obtain a more "macroscopic" Schroedinger equation that is approximate but very useful. Let us see this in 1 dimension (1D) first.

We assume that an external electrostatic potential energy $V(x)$ is applied to a 1-dimensional crystal whose eigenstates are the Bloch function $\Phi_{nk}(x)$ where n is the band index and k is the wavevector. Then according to the discussion at the beginning of section 2.5, the wavefunctions $\Psi(x)$ of the electrons are no longer individual Bloch functions but are wavepackets described as a sum (or integral) of Bloch functions (see equation 2.37)

$$\Psi(x)=\sum_n \int \varphi_n(k)\Phi_{nk}(x)dk \qquad 2.73$$

where the $\varphi_n(k)$ in 2.73 are the weighing factors A(k) of equation 2.37. The wavefunctions $\Psi(x)$ will be eigenstates of the combined hamiltonian of the unperturbed crystal and the applied potential energy $V(x)$

$$\left[H_{cr}(x)+V(x)\right]\Psi(x)=E\Psi(x) \qquad 2.74$$

Several approximations can be made to equation 2.73. To begin, we may assume that interband transitions are not present and the electrons remain in the same band, then the summation over the band index n can be dropped and each band considered separately. Secondly, we may assume that the periodic part $u_{n,k}(x)$ of the Bloch functions does not vary much with k so that we may replace all u_{nk} with the u_{n0}, the function at the band extremum $k=0$. Therefore, we have for the Bloch functions

$$\Phi_{nk}(x)=u_{nk}(x)e^{ikx}\approx u_{no}e^{ikx} \qquad 2.75$$

and for the wavefunction

$$\Psi(x)\approx u_{no}(x)\int \varphi_n(k)e^{ikx}dk=u_{no}(x)\tilde{\varphi}_n(x) \qquad 2.76$$

where $\tilde{\varphi}_n(x)$ is the Fourier transformed $\varphi_n(k)$. To be able to make such an approximation, the external potential must vary slowly. A quantification of this statement will come later. A simple interpretation of this statement is that $V(x)$ can be regarded constant over several atomic sites.

Substituting 2.76 in the Schroedinger equation 2.74 we get

$$H_{cr}\Psi(x)+V(x)u_{no}(x)\tilde{\varphi}_n(x)=Eu_{no}(x)\tilde{\varphi}_n(x) \qquad 2.77$$

The first term of 2.77 can be written

$$H_{cr}(x)\Psi(x)=\int \varphi_n(k)H_{cr}\Phi_{nk}dk=$$

$$=\int \varphi_n(k)E_n(k)\Phi_{nk}(x)dk \qquad 2.78$$

where in 2.78 we have used the fact that the Bloch functions Φ_{nk} are eigenfunctions of H_{cr}. Using our approximation 2.75, equation 2.78 becomes

$$H_{cr}\Psi(x)=u_{no}(x)\int \varphi_n(k)E_n(k)e^{ikx}dk \qquad 2.79$$

Let us now make use of the effective mass approximation for $E_n(k)$. Then 2.79 becomes (remember we are still within the 1D model)

$$H_{cr}\Psi(x)=u_{no}(x)\int \varphi_n(k)\left(E_C + \frac{\hbar^2 k^2}{2m_n^*} \right)e^{ikx}dk$$

The integral of the RHS of the above equation is an inverse Fourier transform. Remembering the properties of this transform—that k becomes $\dfrac{\partial}{\partial x}$ in real space—we get

$$H_{cr}\Psi(x)=u_{no}(x)\left[E_C\tilde{\varphi}_n(x) - \frac{\hbar^2}{2m_n^*}\frac{d^2\tilde{\varphi}_n(x)}{dx^2} \right] \qquad 2.80$$

Substituting 2.80 in 2.77 we get

$$\left[-\frac{\hbar^2}{m_n^*}\frac{d^2}{dx^2}+V(x) \right]\tilde{\varphi}_n(x)=(E-E_C)\tilde{\varphi}_n(x) \qquad 2.81$$

Equation 2.81 is called the effective mass equation. It uses the effective mass approximation for the $E_n(k)$ to obtain an equation that has the exact form of a Schroedinger equation with energy eigenvalues measured from the band edge E_C. The eigenstates of the corresponding operator in 2.81 are not the actual wavefunctions $\Psi(x)$ but are the envelope functions $\tilde{\varphi}_n(x)$ (see figure 2.18), which constitute all the information we want. We are not really interested in the variation of $\Psi(x)$ from site to site, but we are interested in the variation over many sites as this information will reveal either the propagation of an electron through the crystal or the localization of it in a certain portion of the crystal.

The effective mass equation is very useful when we have sandwiches of materials stuck together (which is what devices are made of). Imagine a piece of GaAs of nanometric length L (say $L=2nm$) sandwiched between two pieces of GaAlAs of much greater length.

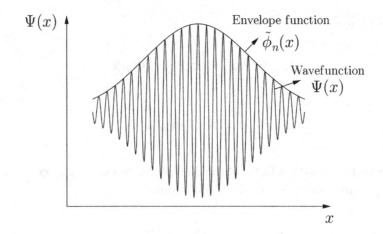

FIGURE 2.18 Real and envelope wavefunction in a semiconductor.

Given that GaAlAs has a much higher E_g, it looks as if the electrons in the conduction band of GaAs cannot move into GaAlAs either to the left or to the right, see figure 2.19. We can use the effective mass equation to calculate the energies of this quantum well. What we have called external potential in the lines that lead to the derivation of the effective mass equation need not be a potential energy supplied by a user but it can be any perturbation to the crystal. In this case we can draw a diagram, see figure 2.20, of the band gap variation along the length of the sandwich.

The conduction band difference, or offset, $E_C(\text{GaAlAs}) - E_C(\text{GaAs}) = \Delta E_C$ can be thought as a quantum well depth that keeps the conduction band electrons of GaAs in GaAs. Likewise for the holes in the valence band. We can therefore write

$$\left[\frac{-\hbar^2}{2m_n^*}\frac{\partial^2}{\partial x^2} - \Delta E_C\right]\tilde{\phi}_n(x) = (E_n - E_C)\tilde{\phi}_n(x) \qquad 2.82$$

where ΔE_C is of course a nonzero constant over only the distance L (= $2nm$) that GaAs extends. Equation 2.82 is of the form of a quantum well (see chapter 1), albeit of a finite

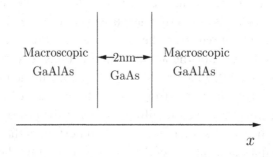

FIGURE 2.19   Schematic picture of a nanosystem in a one dimension extending infinitely in the other two dimensions.

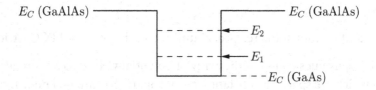

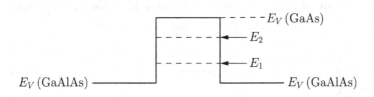

FIGURE 2.20 Band edges and energy levels of the confined nanosystem of figure 2.19.

barrier height, and therefore it will give a set of discrete levels which electrons will occupy (see figure 2.20), calculated relative to E_C. Note that E_C denotes the conduction band edge of a macroscopic piece of GaAs. If L is of the order of microns, all the E_n will almost coincide with E_C, but if L is nanometric, as we have assumed, the $(E_n - E_C)$ in equation 2.82 will be positive and substantial. Confinement of carriers creates what we call quantization. In subsequent chapters we will encounter many such cases in our study of devices.

In three dimensions 2.82 is generalized to

$$\left[\frac{-\hbar^2}{2m_n^*}\nabla^2 + V(r)\right]\tilde{\varphi}_n(r) = (E_n - E_C)\tilde{\varphi}_n(r) \qquad 2.83$$

and for holes in the valence band

$$\left[\frac{-\hbar^2}{2m_p^*}\nabla^2 + V(r)\right]\tilde{\varphi}_n(r) = (E_V - E_n)\tilde{\varphi}_n(r) \qquad 2.84$$

The above equations hold for semiconductors in which E_C and E_V are singly degenerate, i.e. a simple k point is associated with the band extremum. Such are, for example, both E_C and E_V in GaAs or only E_V in Si. On the other hand, we have seen that the conduction band-edge E_C in Si is six-fold degenerate. In this case we need to ascribe a valley index j to the wavefunction $\tilde{\varphi}_n$, i.e. write $\tilde{\varphi}_{n,j}$ and use the appropriate effective mass m_i^* for each direction (i) in each of the 6 valleys at which the conduction band extremum is located. Furthermore, if the effective mass approximation is to be used at an interface of two different semiconductors further complications exist and a further modification is required. All these are described in section 5.4.

PROBLEMS

2.1 Find the unit vectors of the reciprocal unit cell of the BCC and FCC lattice.

2.2 Find the eigenvalues of a periodic array of potential wells of constant depth –Vo and width a which are spaced a constant length d apart (Kroning–Penney model).

2.3 Find the eigenvalues of a BCC lattice of spacing a with only a 1s orbital on each atom. Expand the solution near the band minimum in terms of the magnitude of the wave-vector k and thus obtain an expression for the effective mass of such a hypothetical solid.

2.4 Find an expression for the Bloch energies E(k) of a cubic lattice with s and p orbitals on each atom.

2.5 Assume that a semiconductor is heated to such a high temperature that the excited electrons moved from the valence band to the conduction band are much more than the electrons from the impurities. The semiconductor then becomes intrinsic. Find that temperature.

II

Theory of Conduction

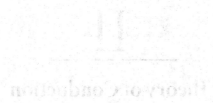

CHAPTER 3

Simple Classical Theory of Conduction

3.1 EXTERNAL VOLTAGES AND FERMI LEVELS

In this chapter we will give an account of the elementary or standard theory of classical conduction, i.e. the theory which is based on the continuity and the drift–diffusion equations. As such, the chapter will deal with the relation between applied voltages and observed currents in both semiconductor materials and devices. A note on notation is mandatory at this stage. In the previous chapters we used the capital letter V for the potential energy of an electron in the Schroedinger equation. In this and subsequent chapters, we would like to reserve this symbol for the quantity voltage or potential, so that the potential energy of an electron corresponding to an arbitrary potential V(x) will be $U(x) = -eV(x) = -1.6 \times 10^{-19}(C)V(x)$. We note that the symbol e is always considered positive.

Imagine a homogeneous piece of metal onto which an external voltage difference V is applied, see figure 3.1a. This voltage will create a potential energy $U(x)$ inside the metal and a 1D electric field $\mathcal{E}$, which, being equal to $\mathcal{E} = -\dfrac{1}{e}\dfrac{dU}{dx}$, will give the rate of change of the potential energy $U(x)$. If the applied electric fields are small compared to the internal electric fields due to the crystalline potential energy V_{cr}, then we can always assume the following two statements:a) that in a small length element dx, $U(x)$ may be considered constant, and b) at the same time within dx, we have enough atoms to form a crystal with all the properties of a crystal that we have discussed so far. Under these circumstances, we can think that the top of the band which is at E_F follows $U(x)$ under the action of the applied V, see figure 3.1b. Therefore under the action of V (or $\mathcal{E}$) the electrons will slide down from the higher energies to the lower energies.

Furthermore, the Fermi level E_F will no longer be constant but the difference in E_F at the two ends of the metal piece will be equal to eV. If we call E_F^l and E_F^r the Fermi levels at the two ends (left and right) of the metal then we will have

$$E_F^l - E_F^r = eV \qquad\qquad 3.1$$

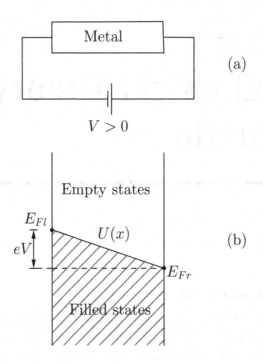

FIGURE 3.1   Bending of the conduction band of a metal with distance x when an external voltage V is applied to it. The external voltage produces an additional internal potential energy U(x) in the metal.

The situation described above has certain underlying assumptions over and above all those just described. First and foremost, the Fermi level is a concept or quantity valid only at equilibrium and pertaining to the whole of the system under consideration. It is a space-independent quantity, unique for any system. One can talk about the Fermi level variation in space if the deviation from equilibrium is weak and the system is large. In principle, one has to solve the Boltzmann equation which gives the probability of occupation of a state in terms of both the position vector r and the wavevector k. We will do that in the next chapter. However, equation 3.1 is always regarded as valid in the sense that the applied electrodes are big enough to provide two thermodynamic reservoirs, one at each end, that keep the Fermi levels constant.

In semiconductors there are further complications due to the existence of two types of carriers, electrons and holes, and the sensitivity of their concentrations on the Fermi level. The well-known relationship

$$np = n_i^2 \text{(equation 2.61 reproduced here)}$$

does not hold away from equilibrium. As we will see, the majority carriers change marginally, while the minority carriers drastically. The concentration of the latter is calculated by means of the continuity equations. However as more of a mathematical device and less as a physical concept, we define the quasi Fermi levels for electrons and holes separately E_{Fn} and E_{Fp}, when usually electrons or holes are each majority carriers respectively, by extending the validity of equations 2.62 and 2.63, i.e. by writing

$$n \equiv n_i exp\left(\frac{E_{Fn} - E_{Fi}}{KT}\right)$$ 3.2a

$$p \equiv n_i exp\left(\frac{E_{Fi} - E_{Fp}}{KT}\right)$$ 3.2b

We emphasize again that equations 3.2a and 3.2b are mere definitions of E_{Fn}, E_{Fp}. Essentially E_{Fn} and E_{Fp} constitute a device by which the functional form for n and p at equilibrium is preserved at non-equilibrium. Multiplying 3.2a by 3.2b we get

$$np = n_i^2 exp\left(\frac{E_{Fn} - E_{Fp}}{KT}\right)$$ 3.3

If we have thermodynamic equilibrium, then there is only one Fermi level, $E_{Fn} = E_{Fp}$ and we get back to the well known law of mass action.

3.2 COLLISIONS AND DRIFT MOBILITY

In section 2.5, we made it clear that a full band cannot exhibit any conductivity because the sum of all the velocities is zero, cf. equation 2.40 and discussion therein. This statement simply states that in a full band there are always as many velocities in a direction as in the opposite. This statement holds true also in a non-full band under the absence of an external field. An external electric field $\mathcal{E}$ can change this balance so that more electrons flow in the opposite direction to $\mathcal{E}$ than in the same direction and the mechanisms to do this are the collisions. (Note that due to the negative charge of the electron, the force is opposite to $\mathcal{E}$.) What happens is that the collisions in the direction of $\mathcal{E}$ become more frequent. Note that the individual velocity of any electron is still given in the presence of a field by $\nabla_k \left[E(k)\right]$, equal to $\hbar k / m^*$ for a parabolic band. The electric field does not change the individual velocities, only their distribution.

Hence, current is a property of an ensemble of electrons; $\mathcal{E}$ changes the distribution of the velocities in k-space: more electrons are flowing with k vectors opposite to the direction of $\mathcal{E}$ than electrons with k vectors in the same direction as $\mathcal{E}$, see figures 3.2a and 3.2b. The average velocity of the excess electrons in the opposite direction to $\mathcal{E}$ is called drift velocity. Although figures 3.2a and 3.2b refer to a 1-dimensional picture of a crystalline solid, they have all the physical ingredients that we will need in this chapter. In the next chapter where more advanced physical models will be used, a 3-dimensional picture will be employed. It is with this aim that we have used vector notation so far here.

The distribution of velocities shown in figure 3.2b is time independent but derives from a highly dynamic situation: electrons are constantly accelerated and scattered by collisions to lower energies. An electron with momentum of, for example, $\hbar k_1$ in 1 dimension is accelerated

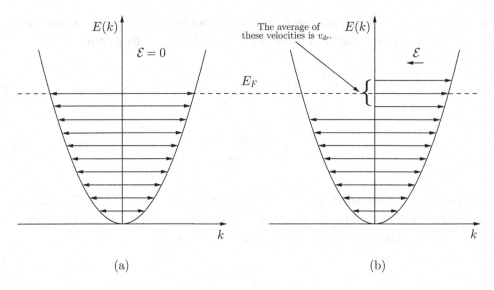

FIGURE 3.2   Distribution of velocities in a conduction band (a) before and (b) after an electric field is applied to a solid.

by the electric field $\mathcal{E}$ to $\hbar k_2 > \hbar k_1$ and then scattered by collision to a state with momentum $\hbar k_3 < \hbar k_2$. As a result of all these processes, one gets the static picture of figure 3.2b with a net collective velocity. In section 2.6, we showed that under the assumptions of a parabolic band and a slowly varying field, an electron's motion can be considered as classical with the effective mass substituted for the real mass. Hence we will consider a set of N classical particles, each with an effective mass m^* and derive the current density J of this system in terms of the electric field $\mathcal{E}$ and the properties of the system, such as the conductivity σ and mobility μ defined by $J = \sigma\mathcal{E}$ and $\sigma = ne\mu$. We emphasize that what follows is the lowest order of an approximate theory of charge transport in semiconductors valid only under the conditions we have described above. More advanced theories will follow.

For such a system of classical charged particles, each with charge $-e$, the total momentum increase $(dP)_{el}$ in time dt due to the electric field is

$$(dP)_{el} = -Ne\mathcal{E}dt \qquad\qquad 3.4$$

At the same time, the system loses momentum due to collisions. The simplest approximation we can make is that each particle loses by collision all the kinetic energy it has gained so that, in our 1-dimensional model, the variation of velocity with time of a single particle will be the one shown in figure 3.3. In this figure the change in time of the velocity of two electrons is shown for clarity, but we can imagine many more electrons all colliding randomly with the lattice at a random frequency that is sharply peaked around a mean value which can be written as $1/\tau$ where τ is the so-called relaxation time and can be thought of as roughly the mean time between collisions with the lattice. At the moment the reader should interpret the word "lattice" very loosely, since we have not specified with what exactly the electrons collide. We leave this critical issue for the next section.

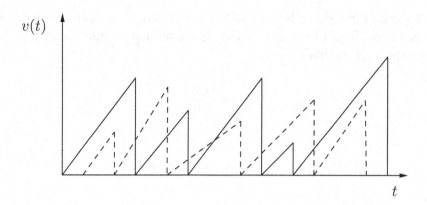

FIGURE 3.3   Time evolution of the velocities of two particles under the action of an electric field and of collisions.

So if the system loses all the momentum it has gained in time τ, in time dt it loses

$$(dP)_{col} = -\frac{P \cdot dt}{\tau}$$ 3.5

Therefore, in a time independent state we must have

$$-Ne\mathcal{E} - \frac{P}{\tau} = 0$$

$$\Rightarrow P = -Ne\mathcal{E}\tau$$ 3.6

But the drift velocity, as explained so far, is the mean of the extra momentum per particle divided by the (effective) mass. Therefore

$$v_{dr} = \frac{P}{N} \times \frac{1}{m^*} = \frac{\langle p \rangle}{m^*}$$

where $\langle p \rangle$ is the average additional momentum per electron. By the word 'additional,' we mean the additional momentum gained by the system due to the electric field.

Using equation 3.6, the above equation can be written as

$$v_{dr} = \frac{e\tau\mathcal{E}}{m^*}$$ 3.7

and hence the mobility μ defined from the relation $v_{dr} = \mu\mathcal{E}$ is given by

$$\mu = \frac{e\tau}{m^*}$$ 3.8

The above classical model can be applied to both electrons in the conduction band and holes in the valence band of a semiconductor. Using the superscripts n and p for electrons and holes respectively we have

$$v_{dr}^n = \frac{e\tau_n \mathcal{E}}{m_n^*} \qquad \text{3.9a}$$

$$v_{dr}^p = \frac{e\tau_p \mathcal{E}}{m_p^*} \qquad \text{3.9b}$$

and for the corresponding mobilities

$$\mu_n = \frac{e\tau_n}{m_n^*} \qquad \text{3.10a}$$

$$\mu_p = \frac{e\tau_p}{m_p^*} \qquad \text{3.10b}$$

3.3 MECHANISMS OF SCATTERING

So far we have talked about collisions and scattering in more or less loose terms, without specifying the source of the scattering, or equivalently, with what exactly the electrons collide. From early undergraduate years, the student learns that electrons collide with impurities and the lattice vibrations. Here we want to allocate a few lines to clarify a few points. In chapter 2, we proved that the eigenstates of electrons in a periodic lattice are Bloch functions, i.e. modulated plane waves, each eigenstate corresponding to an energy $E_n(\textbf{k})$, where $\textbf{k}$ is a wavevector and n is a band index. The wavevector $\textbf{k}$ acts like the macroscopic momentum $\textbf{p}$ of an electron, i.e. the two are related by $\textbf{p} = \hbar\textbf{k}$. Furthermore, $\mathcal{E}$ is given by the time derivative of $\textbf{k}$. In the 1 dimension we are considering here

$$e\mathcal{E} = \hbar\frac{dk}{dt} \text{ (2.41 reproduced here)}$$

So when a field is applied, the electrons will move to higher and higher k vectors, that is to higher and higher momenta in accordance with 2.41. It seems that nothing will stop this process of acceleration. Obviously, the acceleration will be stopped, as assumed in the previous section, but what will stop the validity of 2.41 or equivalently create collisions? The answer is: anything that will lift or destroy the periodicity of the crystalline potential which gives rise to the Bloch functions and consequently the validity of 2.41. These are:

a. lattice vibrations

b. impurities

c. crystal defects (like vacant sites)

d. interfaces

Whereas b), c), d) are obvious deviations from periodicity, as far as a) we note that as the atoms of the crystal vibrate they move away from their ideal static positions and therefore vibrations indeed constitute deviations from periodicity. The frequency of the electron collisions with the vibrations will depend on the exact type of these vibrations. An introduction into the formal theory of lattice vibrations is given in Appendix B where the reader will find a complete classical theory. The interactions of the electrons with the perturbing potentials caused by a, b, c, d above constitute a major portion of solid state physics. Luckily we do not need this theory itself to formulate a theory of electronic conduction, we only need the results, i.e. the time rate of collisions which are given by the relaxation times τ.

Each scattering mechanism has its own relaxation time, so how come we have used a simple relaxation time in our formulation of mobility? The following argument shows how this is possible. If τ_i is the relaxation time of the ith scattering mechanism, then $1/\tau_i$ is the probability of scattering per unit time of the ith mechanism. The total probability of scattering per unit time τ by any mechanism is then the sum of the individual ones assuming that each mechanism is independent of the others. Then

$$\frac{1}{\tau} = \sum_i \frac{1}{\tau_i} \qquad\qquad 3.11$$

However, each of the τ_i exhibits a different temperature dependence. When the temperature rises, it is reasonable to suggest that the amplitude of the lattice vibrations increase, irrespective of their type, so we expect the collision of electrons with these vibrations to become more frequent. A lengthy calculation shows that the relaxation time for scattering with the lattice vibrations (or phonons) is

$$\tau_{ph} \propto T^{-3/2} \qquad\qquad 3.12$$

Conversely, the collisions of electrons with charged impurities are expected to become less frequent as the temperature rises because the electrons have a higher energy at a higher temperature and hence are expected to overcome the attraction or repulsion of charged impurities. A simpler calculation this time shows

$$\tau_{imp} \propto T^{+3/2} \qquad\qquad 3.13$$

If we add the two mechanisms appropriately, the total relaxation time τ then exhibits a maximum at a certain temperature as figure 3.4 shows.

3.4 RECOMBINATION OF CARRIERS

In the previous section we examined the effects of the application of an external electric field to a metal or a semiconductor. This is the most common way to deviate from thermodynamic equilibrium. But there are other ways, notably the application of an electromagnetic field (light) and the injection of carriers from one side of the sample to the other one.

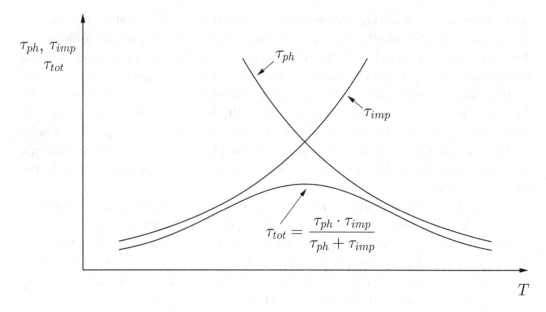

FIGURE 3.4 Composite relaxation time and relaxation times for impurities τ_{imp} and lattice vibrations τ_{ph} as a function of temperature T.

The former creates electron–hole pairs, whereas the latter simply adds more electrons or holes into the system, thereby destroying the relation

$$np = n_i^2 \quad (2.61 \text{ reproduced here})$$

that we have seen in section 2.7. Key to understanding both processes is the concept of recombination of electrons with holes in semiconductors. The theory we are going to give is adequate, provided that the condition of "low injection" holds, which means that the majority carriers remain almost unaltered (percentagewise), whereas it is the minority carriers that change significantly.

Let us take an arithmetic example to see how this is possible. Consider a sample of Si doped with $N_D = 10^{15}/cm^3$ donors. Since the intrinsic concentration of Si $n_i \approx 10^{10}/cm^3$ at room temperature we deduce that

$$n = N_D = 10^{15}/cm^3 \, p = 10^5/cm^3$$

Now imagine an injection of $10^{10}/cm^3$ of holes. This will change the number of holes tremendously (by 5 orders of magnitude), but even if we assume that all these holes recombine with electrons, the precentage change in the number of electrons is insignificant. Therefore in what follows we will assume that the number of majority carriers remains essentially unaltered and the process of injection or the incidence of light do not change the character of the semiconductor, i.e. change it from N to P and vice versa.

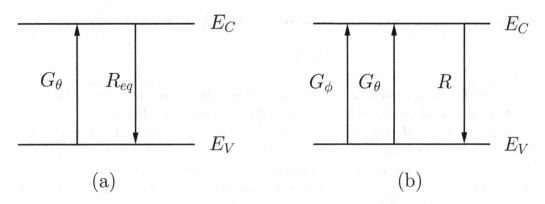

FIGURE 3.5 (a) Arrows G_θ and R_{eq} portray the thermal generation and recombination of carriers respectively between the valence and conduction bands of a semiconductor at equilibrium, (b) a non-equilibrium case with G_φ indicating the excitation of electrons from the valence band to the conduction band due to an electromagnetic field of the appropriate frequency.

Of course the idea that we have just expressed that all injected or photogenerated minority carriers will recombine is not representative of reality and was only used to show that the majority carrier concentration remains essentially unaltered. Both the photo-generation of carriers and the injection of minority carriers create a new state of things that is not given by equation 2.61. This new state forms the basis for the understanding of the PN junction, which is the basic building block of both the bipolar junction transistor and the metal–oxide–semiconductor field effect transistor.

Let us begin with an N-type semiconductor in thermodynamic equilibrium. Under this condition, equation 2.61 holds true, but it must be interpreted as a case of dynamic equilibrium in which electrons constantly recombine with holes and new electrons in equal amounts are created thermally. These two processes are shown schematically in figure 3.5a by the corresponding arrows connecting the band edges E_C, E_V. Since we have an equilibrium, i.e. a time-independent set of variables we must have (the subscript o denoting the equilibrium)

$$\frac{dn_0}{dt} = G_\theta - R_{eq} = 0 \qquad\qquad 3.14$$

where G_θ is the rate of thermally activated electrons (producing holes) and R_{eq} is the rate of electron recombination in equilibrium. Therefore, as expected,

$$G_\theta = R_{eq} \qquad\qquad 3.15$$

G_θ is determined solely by temperature, whereas the recombination rate R is proportional to both n and p. Hence we can write quite generally (not only for thermodynamic equilibrium)

$$R = C \cdot n \cdot p \qquad\qquad 3.16$$

where C is a constant. For the case exclusively of thermodynamic equilibrium we get

$$G_\theta = R_{eq} = C \cdot n_0 \cdot p_0 \qquad\qquad 3.17$$

Now consider as an example of a deviation from thermodynamic equilibrium, an electromagnetic field incident on the N-type semiconductor. This field will excite electrons from the valence band to the conduction band creating electron–hole pairs at a constant rate G_φ as shown in figure 3.5b. We no longer have a time-independent situation as light keeps piling up electrons in the conduction band creating holes in the valence band. How will a new steady state (not a thermodynamic equilibrium) be reached if electron–hole pairs keep being created? Answer: recombination will increase.

The net thermal generation minus recombination rate at non-equilibrium is

$$G_\theta - R = G_\theta - Cn_0\left(p_0 + \Delta p\right) \qquad\qquad 3.18$$

where we have used the fact that the number of majority carriers is not appreciably changed. But given equation 3.17, for the net rate of whole recombination away from equilibrium we have

$$G_\theta - R = -Cn_0\Delta p = \frac{-\Delta p}{\tau_p} \qquad\qquad 3.19$$

The quantity τ_p is called the recombination time and although the symbol is the same, it is not the same as the relaxation time for collisions which usually bears the same symbol.

Then to get the net rate of creation of electron–hole pairs, we must subtract from G_φ the above net rate of recombination

$$\frac{dp_n}{dt} = G_\varphi - \frac{\Delta p_n}{\tau_p} \Rightarrow \frac{d\left(\Delta p_n\right)}{dt} = G_\varphi - \frac{\Delta p_n}{\tau_p} \qquad\qquad 3.20$$

where in equation 3.20 we have included a subscript n to indicate that the equation refers to holes in an N-type semiconductor. This will constitute a standard notation from this point onwards. From the above equation we can see that the incident light (of $\hbar\omega \geqslant E_g$) will create a new steady state after a time much longer than τ_p, where the number of holes would have increased by

$$\Delta p_n = G_\varphi \tau_p \qquad\qquad 3.21$$

In fact, the complete solution of 3.20 (a 1st order differential equation) under the initial condition that $\Delta p_n(0) = 0$ is

$$\Delta p_n(t) = G_\varphi \tau_p \left(1 - e^{-t/\tau_p}\right) \qquad\qquad 3.22$$

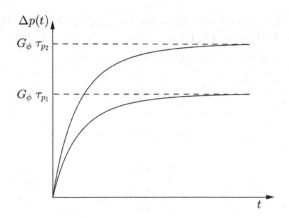

FIGURE 3.6 Time evolution of the excess holes $\Delta p(t)$ after an electromagnetic field has been switched on.

The graph of this function is shown in figure 3.6 for two values of τ_p. For a P-type semiconductor, by exactly the same kind of arguments we get

$$\frac{d\Delta n_p(t)}{dt} = G_\varphi - \frac{\Delta n_p}{\tau_n}$$

and for its solution

$$\Delta n_p(t) = G_\varphi \tau_n \left(1 - e^{-t/\tau_n}\right) \qquad 3.23$$

We have so far considered generation and recombination processes which consist of only transitions from the top of the valence band E_V to the bottom of the conduction band E_C and vice versa. However, there are other indirect and more intricate generation and recombination processes that occur through what we call "states in the gap" of the semiconductor. These are states in the gap which derive from unwanted impurity atoms in the semiconductor just like the As and Ga atoms which are intentionally implanted in Si to create donors and acceptors respectively. These impurities are usually metal atoms that are left over after the purification process of the semiconductor.

Let us call E_t the energy of one such state. Let us also distinguish whether the energy state in the gap is occupied or empty by the symbols $E_t(\text{occ})$ and $E_t(\text{emp})$ respectively. Then the following 4 electron processes may occur

1. $E_C \rightarrow E_t(\text{emp})$ that is an electron capture

2. $E_t(\text{occ}) \rightarrow E_C$ that is an electron emission

3. $E_t(\text{occ}) \rightarrow E_V$ that is a hole capture

4. $E_V \rightarrow E_t(\text{emp})$ that is hole emission

At steady state the capture (cap) and emission (em) processes must balance out for each type of carrier separately. Then in a notation with capital R standing for rate and subscripts n, p standing for electrons and holes respectively we must have

$$R_n^{cap} = R_n^{em}$$

and

$$R_p^{em} = R_p^{cap}$$

This problem was tackled by Shockley, Read, and Hall and hence bears their names. Under certain simplifying assumptions the recombination rate R_t of electrons and holes through a state E_t whose energy is equal to the mid-gap energy is

$$R_t = \frac{np}{\tau_n n + \tau_p p} \qquad 3.24$$

where τ_n and τ_p are constants with the dimension of time.

3.5 DIFFUSION CURRENT

Equation 3.20 and its solution, equation 3.22, assume that the excess carriers $\Delta p(t)$ have no space-dependence. However, as we shall see in the case of the pn junction in this chapter, the injection of minority carriers occurs at a given direction and our current assumption of no space dependence for Δp is not valid. In this case, when the excess carriers have a non-uniform distribution in space, we encounter the phenomenon of carrier diffusion.

Figure 3.7 shows an arbitrary density of holes that is x dependent (1-D model). At each point x there will be a hole current J_p^{diff}, called a diffusion current, which will be sent in the direction of decreasing $p(x)$ and which will be given by

$$J_p^{diff} = -eD_p \frac{dp}{dx} \qquad 3.25$$

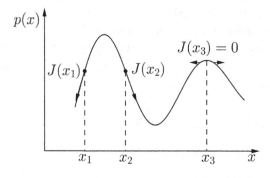

FIGURE 3.7   A variation of a hole density in space gives rise to a space dependent diffusion current. The arrows indicate only the direction of flow.

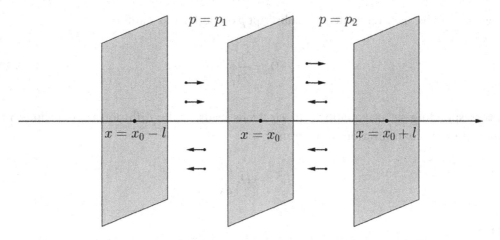

FIGURE 3.8 Geometry for the calculation of the diffusion current.

where D_p is a constant called the diffusion coefficient for holes. The arrows in figure 3.7 along the tangent of the $p(x)$ curve indicate simply the direction of the current density J_p^{diff}, left or right. This is called Fick's first law of diffusion and it can be proved as follows.

Imagine a plane at $x = x_0$, vertical to the direction x, of finite area A and two other planes, also vertical to x and of area A at a distance l of either side of the plane at x_0, see figure 3.8. Now the distance l is not arbitrary, but it is specifically chosen to be the distance the carriers travel without being scattered. It is called the mean-free path. Hence it is related to the relaxation time τ of holes that we have encountered in our elementary theory of transport by the formula

$$l = v_F \tau \qquad\qquad 3.26$$

where v_F is the velocity of holes near the Fermi level. Note that we have dropped the subscript p on τ. Therefore during the time τ, half of the holes enclosed between the planes at x_0 and at $x_0 - l$ will move to the right and half of the holes between x_0 and $x_0 + l$ will move to the left. This is due to the carrier's random (brownian) motion. The mean free path l is usually a few nanometers so that the concentration p can be considered constant between x_0 and $x_0 - l$, say equal to p_1 and between x_0 and $x_0 + l$, equal to p_2.

Then the current crossing the plane at x_0 is

$$J_p^{diff} = \frac{e\left[\dfrac{1}{2}p_1 Al - \dfrac{1}{2}p_2 Al\right]}{A\tau} = \frac{-el}{2\tau}\left(p_2 - p_1\right) \qquad\qquad 3.27$$

Now $p_2 - p_1 = (dp/dx)l$ and hence

$$J_p^{diff} = \frac{-l_p^2}{2\tau_p}\frac{dp}{dx} = -eD_p\frac{dp}{dx} \qquad\qquad 3.28$$

where in 3.28 we have reinstated the subscript p on both l and τ and

$$D_p = \frac{l_p^2}{2\tau_p} \qquad\qquad 3.29$$

A corresponding set of equations for electrons leads to the diffusion current due to a gradient in the concentration of electrons

$$J_n^{diff} = eD_n \frac{dn}{dx} \qquad\qquad 3.30$$

where

$$D_n = \frac{l_n^2}{2\tau_n} \qquad\qquad 3.31$$

with l_n and τ_n the mean free path and relaxation time of an electron respectively. The absence of the minus sign in 3.30, when compared to 3.28, is due to the negative charge of the electron as opposed to that of a hole.

3.6 CONTINUITY EQUATIONS

In section 3.4 we dealt with a temporal change in the concentration of minority carriers and in section 3.5 with space-dependent concentrations. But in most cases, for example, when a steady state has not settled in, we have both time and space-dependent concentrations of carriers. How do we deal with that? The answer is through the continuity equations for the preservation of particles and correspondingly of their charge. The classical continuity equation of electromagnetism states, in 1 dimension,

$$\frac{d\rho}{dt} = -\frac{\partial J}{\partial x} \qquad\qquad 3.32a$$

and in 3 dimensions (3D)

$$\frac{d\rho}{dt} = -\nabla J \qquad\qquad 3.32b$$

where ρ is the charge density and J is the current density. The physical meaning of the above equations is that if the current density going into an infinitesimal volume at x between time t and $t+dt$ is more than the current going out of that volume at $x+dx$ during the same time interval (i.e. there is a gradient in the current density J) then some charge density $d\rho$ has been left behind in the infinitesimal volume during the time interval dt, see figure 3.9.

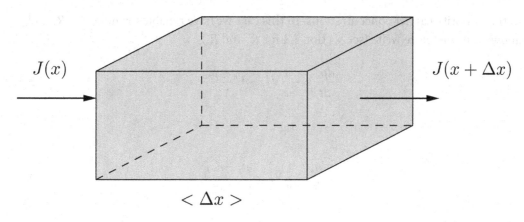

FIGURE 3.9   A current enters normally an infinitesimal volume at x and leaves at x + Δx during an infinitesimal time Δt. What is left in the volume constitutes the local increase of change density in time Δt.

In semiconductors we are not interested only in the total charge density ρ but we require more detailed information: we want to know the rates of change of the electrons and the holes separately. Following the same argument as in the previous paragraph, we can examine the rates of change in an infinitesimal volume of electrons and holes separately, $\dfrac{dn}{dt}$ and $\dfrac{dp}{dt}$ respectively, and state that for each type of carrier concentration (generalizing 3.32a and 3.32b)

$$\begin{matrix} \text{rate of change} \\ \text{of concentration} \end{matrix} = \begin{matrix} \text{gradient of} \\ \text{corresponding} \\ \text{current} \end{matrix} - \begin{matrix} \text{rate of} \\ \text{recombination} \end{matrix}$$

As a numerical example of the above statement, if in a microsecond 100 electrons go into a volume and 70 come out of this volume (in a microsecond) while 10 have recombined with holes, the rate of increase is 20 electrons per microsecond.

We have to express the above relation in strict mathematical language. For holes, taking into account that a concentration p is equivalent to a charge $\rho = ep$, we get (in 3D)

$$\frac{dp_n}{dt} = -\frac{1}{e}\nabla J_p - R_p \qquad\qquad 3.33$$

and correspondingly for electrons

$$\frac{dn_p}{dt} = \frac{1}{e}\nabla J_n - R_n \qquad\qquad 3.34$$

where R_n, R_p are the recombination rates of electrons and holes. Note the change of sign in 3.34 compared to 3.33. As we have explained in section 3.4, the majority carriers are only marginally affected by either photogeneration or injection of carriers, so the interest lies in the minority carrier concentrations. In this case we have reliable formulae for R_n and R_p and we can therefore write (see section 3.4 for R_n and R_p)

$$\frac{dp_n}{dt} = -\frac{1}{e}\nabla J_p - \frac{p_n - p_{n0}}{\tau_p} \qquad 3.35$$

and

$$\frac{dn_p}{dt} = \frac{1}{e}\nabla J_n - \frac{n_p - n_{p0}}{\tau_n} \qquad 3.36$$

where n_{p0} and p_{n0} are the equilibrium concentrations of electrons and holes respectively.

The following case illustrates the use of the continuity equations in space-dependent problems. Electromagnetic (EM) radiation of frequency ω with $\hbar\omega > E_g$ is incident normally to a heavily doped semi-infinite N-type semiconductor. The EM radiation creates electron–hole pairs in a volume of infinitesimally short length near the surface to which the EM is incident, see figure 3.10. Calculate the concentration of minority carriers as a function of the distance from this surface in a steady state when a steady excess of carriers $\Delta n_p(0) = \Delta p_n(0)$ has been created on the aforementioned surface.

As light provides electron–hole pairs at the surface (the one normal to. say, the z axis, at $z = 0$), the concentration of holes increases at those points. Hence, a diffusion current is created which is directed towards the interior of the solid, see figure 3.10. It is natural therefore to expect that there will come a point when those holes, produced by photogeneration, will replace exactly those holes that are taken away by diffusion.We can ignore the small distance that light has penetrated into the semiconductor (not always) and assume that there is not an electric field in the semiconductor and the only electric current present is due to diffusion.

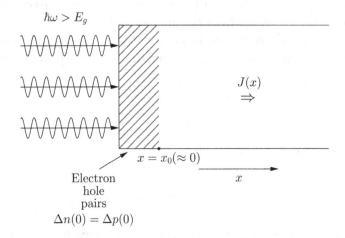

FIGURE 3.10 An incident electromagnetic field changes the electron and hole concentration at one end of a very long bar, thus creating a flow into the semi-infinite bar.

Then by the assumption of a steady state prevailing, we have for the holes, i.e. the minority carriers

$$\frac{dp_n}{dt} = 0$$

and assuming a unidirectional flow due to the normal incidence

$$\frac{d}{dz} J_p^{diff} = -eD_p \frac{d^2 p_n}{dz^2}$$

Then equation 3.35 reduces to

$$0 = D_p \frac{d^2 p_n}{dz^2} - \frac{p_n - p_{n0}}{\tau_p} \qquad 3.37$$

Putting $p_n(z) - p_{n0} = \Delta p_n(z)$ we have

$$0 = \frac{d^2 \Delta p_n(z)}{dz^2} - \frac{\Delta p_n(z)}{D_p \tau_p} = \frac{d^2 \Delta p_n(z)}{dz^2} - \frac{\Delta p_n(z)}{L_p^2}$$

where

$$L_p^2 = D_p \tau_p$$

The general solution of this differential equation is

$$\Delta p_n(z) = A e^{z/L_p} - B e^{-z/L_p}$$

where A and B are constants. The constant A must be zero, $A = 0$, otherwise $\Delta p_n(z)$ is not bounded. Assume that at the surface the deviation from the equilibrium concentration is the constant $\Delta p_n(0)$. Then

$$\Delta p_n(z) = \Delta p_n(0) e^{-z/L_p} \qquad 3.38$$

Although this analysis pertains to photogeneration at a semi-infinite rod it will be shown to describe the physics of the PN junction that we are going to describe next and which is the basic building block of the majority of transistors used at present.

3.7 THE IDEAL PN JUNCTION AT EQUILIBRIUM

We imagine two pieces of doped semiconductor materials, one P-type and one N-type coming close together and forming a PN junction. Of course this is not the way PN junctions are manufactured, but it is a useful thought process in order to understand the real

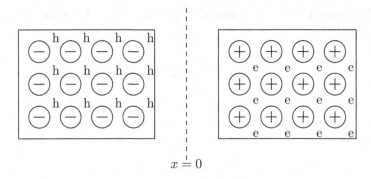

FIGURE 3.11 Schematic carrier and dopant concentrations of a P and an N type semiconductor piece before they are brought into contact to make a PN junction.

processes occurring at the interface of the P- and N-type materials. Figure 3.11 portrays the situation when the two pieces have not touched each other and figure 3.12 gives the situation when a PN junction has been formed and an equilibrium has been achieved, as we show below.

When the two neutral semiconducting pieces make contact with each other, electrons from the N-type semiconductor will diffuse to the P-type (because the concentration of electrons is much lower there) and likewise holes from the P-type semiconductor will diffuse towards the N-type (because the concentration of holes is much lower there). On their way, the two types of carriers will annihilate themselves by recombination, leaving near the surface the charged impurities of each side (portrayed as circles in figure 3.11) without a corresponding neutralizing charge, i.e. a space charge region will be established, also called depletion layer, as shown in figure 3.12. The layer is called the depletion layer because the concentration of both electrons and holes is much reduced there.

The positively charged donors in the N-type semiconductor and the negatively charged acceptors in the P-type semiconductor near the surface create an electric field $\mathcal{E}$ whose direction opposes the further diffusion of electrons and holes. Note that the concentration gradients of both electrons and holes does not become zero anywhere inside the depletion region but a dynamic steady state is reached: the "built in" electric field $\mathcal{E}$ is creating a current density in the opposite direction of the diffusion current that keeps the concentration gradients

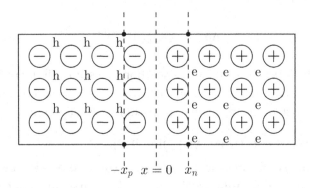

FIGURE 3.12 Schematic carrier and dopant concentrations after the junction has been formed.

constant. Note also that this situation is an equilibrium situation since no energy enters or leaves the system. The above description pertains to the internal electronic density structure before any voltage is applied to the junction.

Before we give a microscopic view of the charge flow in the PN junction, we deal first with the electrostatics of the junction. Certain simplifying assumptions, which are reasonably accurate, are needed in order to get closed expressions for the potential and the electric field built in the diode. We assume that complete annihilation of the mobile charge carriers n and p has occurred inside the space charge–depletion layer, and the charge density, there, is given by the respective doping concentrations (with the right sign) instead of the more general expression $\rho = e\left(N_d^+ + p - N_A^- - n\right)$. Furthermore, we assume that this annihilation has abrupt limits. Further assumptions will be required for the PN junction under bias. For the time, we assume that the depletion layer extends from $-x_p$ to 0 in the P side and from 0 to x_n in the N side of the juction. The charge density ρ in the limit of an abrupt junction is shown in figure 3.13a. Therefore, Poisson's equation for this 1-D problem inside the depletion layer becomes

$$\frac{d^2V}{dx^2} = \begin{cases} +\dfrac{eN_A^-}{\varepsilon_s}, & -x_p \leqslant x < 0 \\[4mm] -\dfrac{eN_D^+}{\varepsilon_s}, & 0 < x \leqslant x_n \end{cases} \qquad 3.39$$

where, as noted, x_p and x_n are the lengths of the P and N part of the PN junction respectively. We have also assumed that the dielectric constant ε_s of the semiconductor has not been affected by the formation of the depletion layer. We note that the semiconductor junction was initially neutral and the space charge layer has appeared as a result of mutual annihilation of electrons and holes. Therefore, the total charge has been preserved and we have

$$N_D^+ x_n = N_A^- x_p \qquad 3.40$$

By the definition of x_n and x_p, we know the length of the depletion region (also called width sometimes) W is

$$W = x_p + x_n \qquad 3.41$$

The boundary conditions for the simple differential equation 3.39 are that the electric field $\mathcal{E} = -\dfrac{dV}{dx}$ is zero outside the depletion region. Therefore $\mathcal{E}\left(x_p\right) = \mathcal{E}\left(x_n\right) = 0$. We furthermore denote by $V_{bi} = V(x_n) - V\left(-x_p\right)$ the built-in potential difference inside the space charge layer or depletion layer. Integrating 3.39 once and using the boundary conditions we get

$$\mathcal{E}(x) = -\frac{dV}{dx} = \frac{-eN_A^-\left(x + x_p\right)}{\varepsilon_s}, \text{ for } -x_p \leqslant x \leqslant 0 \qquad 3.42a$$

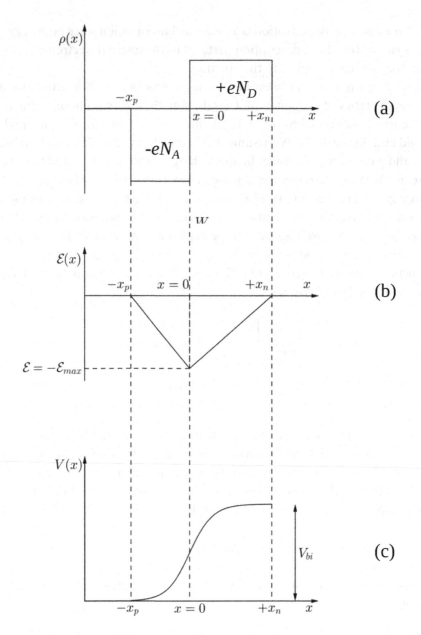

FIGURE 3.13   Change density (a), electric field (b), and potential (c) along the length of a PN junction.

and

$$\mathcal{E}(x) = -\frac{dV}{dx} = \frac{eN_D^+(x-x_n)}{\varepsilon_s}, \text{ for } 0 \leqslant x \leqslant x_n \qquad 3.42b$$

Note that at $x = 0$, equations 3.42a and 3.42b must give the same value. This value is easily seen to be the maximum of the absolute value of the electric field which we denote as $\mathcal{E}_{max}$. The graph of $\mathcal{E}(x)$ is shown in figure 3.13b.

The built-in potential, defined above, is the integral of the electric field, and by simple geometry

$$V_{bi} = -\int_{-x_p}^{x_n} \mathcal{E}(x)\,dx = \frac{1}{2}\mathcal{E}_{max}W \qquad 3.43$$

From equations 3.42a, 3.42b we get

$$\mathcal{E}_{max} = \frac{eN_A^+ x_p}{\varepsilon_s} \qquad 3.44a$$

$$\mathcal{E}_{max} = \frac{eN_D^- x_n}{\varepsilon_s} \qquad 3.44b$$

Hence the length W of the depletion region is

$$W = \frac{\varepsilon_s}{e}\left(\frac{1}{N_A^-} + \frac{1}{N_D^+}\right)\mathcal{E}_{max}$$

and by the use of 3.43 and the fact that at room temperature all impurities are ionized

$$W = \sqrt{\frac{2\varepsilon_s}{e}\left(\frac{1}{N_A} + \frac{1}{N_D}\right)V_{bi}} \qquad 3.45$$

Furthermore, using 3.44a and 3.44b in 3.42a and 3.42b respectively for the electric field we get

$$\mathcal{E}(x) = -\mathcal{E}_{max}\left(1 + \frac{x}{x_p}\right) \text{ for } -x_p \leqslant x \leqslant 0 \qquad 3.46a$$

$$\mathcal{E}(x) = \mathcal{E}_{max}\left(\frac{x}{x_n} - 1\right) \text{ for } 0 \leqslant x \leqslant x_n \qquad 3.46b$$

Note that both the above expressions give $\mathcal{E}(0) = -\mathcal{E}_{max}$ and each gives zero at the respective boundaries $-x_p$, x_n, see figure 3.13b. Integrating once more and choosing $V(-x_p) = 0$ we get

$$V(x) = \mathcal{E}_{max}\left(\frac{x^2}{2x_p} + x\right) + \frac{\mathcal{E}_{max}x_p}{2} \quad -x_p \leqslant x \leqslant 0 \qquad 3.47a$$

and using $V_{bi} = V(x_n) - V(x_p) = V(x_n)$

$$V(x) = -\mathcal{E}_{max}\left(\frac{x^2}{2x_n} - x\right) + V_{bi} - \frac{\mathcal{E}_{max}x_n}{2} \quad 0 \leqslant x \leqslant x_n \qquad 3.47b$$

The graph of $V(x)$ is shown in figure 3.13c.

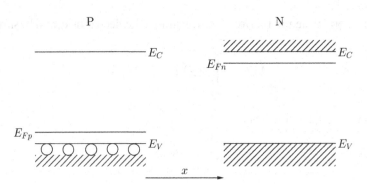

FIGURE 3.14a Position of Fermi levels before contact of the P and N semiconductor pieces.

We now come to a microscopic description of the PN junction, meaning a description of the energy bands and charge flow in the junction. Figure 3.14a reproduces the standard band model of an isolated P-type and an isolated N-type semiconductor. This figure is a partial reproduction of figure 2.14a-b that we have discussed in chapter 2 with a slight modification of notation. The Fermi level in the N-type semiconductor is denoted E_{Fn} and the Fermi level in the P-type E_{Fp}. When the two pieces are brought together to form the junction (a thought experiment as we have emphasized), the two Fermi levels must be equalized since any system under equilibrium has a unique Fermi level. It does not matter if E_{Fp} will come up in energy or E_{Fn} will go down. What matters is that a unique E_F, constant in space must exist. The only way that this can happen is if the bands bend near the interface as figure 3.14b shows.

What will cause the bands to bend? The electrostatic potential energy present in the space charge region or depletion layer constitutes an additional energy that is added to that

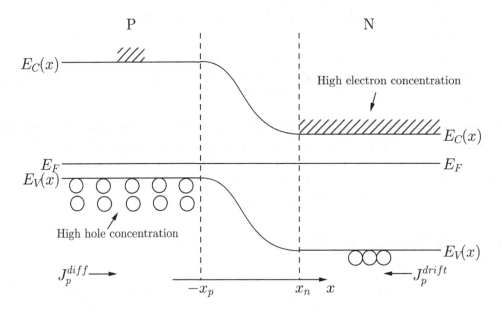

FIGURE 3.14b Band structure of the PN junction at zero bias.

of the crystalline potential energy so that the bending of the bands is a direct result of the built-in potential we have just calculated. It may surprise the reader that the electrostatic potential we have just calculated (see equation 3.47) is monotonically increasing, whereas the bending of the bands is monotonically decreasing. This is only due to the fact that bands denote energies of electrons, whereas the Poisson's equation refers to the potential of a positive test charge. Hence we must multiply the solution of the Poisson equation by (-e) to get the band bending. This relation between band bending and the solution of the Poisson equation derives from the argument we presented in the first section of this chapter regarding voltage and potential energy of electrons.

Under the absence of an external field, the current is zero. This zero current however is the sum of two non-zero currents which add up to zero as we have already mentioned. We will consider each of them in detail using our previous notation for the concentrations of electrons and holes by adding a subscript denoting the region in which the concentration is calculated. Thus n_n denotes the concentration of electrons in the N-type semiconductor side of a PN junction and n_p denotes the concentration of electrons in the P-side side of a PN junction and so on.

We first observe that $n_n \gg n_p$ in the PN junction and hence there will be a diffusion current of electrons from the N-side to the P-side. However this diffusion current is much smaller than what would have been calculated by Fick's law because there is a barrier to the motion of electrons from the N-side to the P-side due to the built-in potential V_{bi}. On the contrary the potential difference V_{bi} acts as a down-hill potential for the electrons coming from the P-side to the N-side. This current is a drift current because electrons move under the action of the electric field producing the V_{bi}. In an obvious notation we therefore write

$$J_n^{drift} + J_n^{diff} = 0 \qquad\qquad 3.48$$

Likewise we observe a concentration difference between p_p and p_n. Therefore, a diffusion current of holes will appear from the P-side to N-side of the PN junction. Again this diffusion current is opposed by the barrier presented by the band bending. At first sight this seems to be a down-hill potential instead of an up-hill one, but we have to remember that we are dealing with holes where as the energies shown are those of electrons, so again the energy gradient shown by the band bending has to be inverted. Also, in analogy with the electrons in the P-side, the holes in the N-type semiconductor will drift by the existing built-in field to the P-type semiconductor side of the junction, as shown by the arrows in figure 3.14b. The two currents just described will cancel each other, so we will have

$$J_p^{drift} + J_p^{diff} = 0 \qquad\qquad 3.49$$

3.8 THE IDEAL PN JUNCTION UNDER BIAS

What mainly happens when a voltage V is applied to the PN junction is that the built-in voltage is changed. Given that the depletion layer is almost empty of mobile charge carriers, and hence has a very high resistivity, we consider that the applied voltage V is dropped

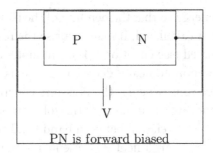

FIGURE 3.15a   Definition of the forward bias in a PN junction.

only across the space charge–depletion layer and not over the entire device length which includes the neutral P and N layers. This statement is going to be slightly modified later. We can view it at the moment as a very good approximation. Before proceeding we need a convention; we will consider the applied voltage to be positive when the positive electrode of the voltage generator is connected to the P-side of the junction, see figure 3.15a. As we can see from figure 3.15b, when $V > 0$ the built-in voltage is reduced to $V_{bi} - V$ and when $V < 0$ it is increased to $V_{bi} + |V|$. Hence, in all cases we can write that the internal barrier becomes $= V_{bi} - V$. To see how this happens we recall from section 3.1 that when a voltage is applied to a specimen, the Fermi level is no longer unique but the Fermi levels at the two ends of the material differ by the applied voltage V. We emphasize that the term "Fermi level" here should be understood within the concept of a quasi–Fermi level discussed in section 3.1. Consider first that $V > 0$. Then the N-side is electrostatically lower than the P-side, which means that the Fermi level at this end of the junction is higher than the P-side (remember always to multiply by $[-e]$ to go from voltages to electronic levels). Then the whole set of electronic levels at the end of the N-side is lifted compared to the neutral part of the P-side, as shown in figure 3.15b, and consequently the barrier for the electrons of the N-type semiconductor to jump to the P-side has been decreased to $V_{bi} - V$. A close examination of figure 3.15b also shows that the barrier for holes has also been reduced

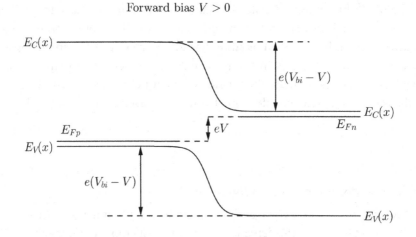

FIGURE 3.15b   Band structure of a PN junction under forward bias.

Reverse bias $V < 0$

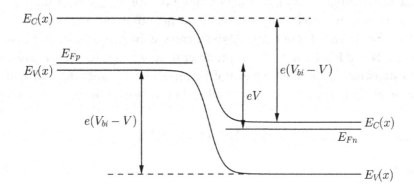

FIGURE 3.15c   Band structure of a PN junction under reverse bias.

accordingly to $V_{bi} - V$. Exactly parallel arguments lead to the band diagram of the PN junction in reverse bias as shown in figure 3.15c.

Once the barrier to jumping across to the other side has been reduced (for electrons in the N-side and holes in the P-side), the diffusion currents J_n^{diff} and J_p^{diff} have increased exponentially with the applied voltage because of an exponentially increased injection, as we will show, of the two types of majority carriers to their opposite side. Note that as the electrons in the N-side are injected into the P-side and the holes in the P-side into the N-side, they change character and become minority carriers, whereas they were majority carriers before the injection. The concentrations n_p and p_n are shown in figure 3.16. These are the key quantities for the understanding of the working of the PN junction. We therefore have to obtain quantitative expressions for them. The problem of the distribution of the injected holes from the P-side into the neutral part of the N-side of the junction is mathematically the same as the example we solved in section 3.6. To see this, note that we have a steady state, it is a 1-D problem and the drift current is zero. The concentration of the holes in the N-side (minority carriers) will therefore be of the form of equation 3.38

$$\Delta p_n(x) = \Delta p_n(x_n) exp\left[-(x-x_n)/L_p\right]$$

3.50

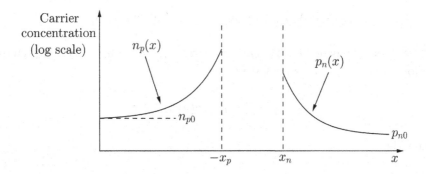

FIGURE 3.16   Minority carrier concentrations across an ideal PN junction.

Since the injection occurs at x_n (and not at $x = 0$), we have changed accordingly the argument of the exponential. A similar equation will hold for electrons in the P-side of the junction. Therefore, we need only evaluate the quantities $\Delta p_n(x_n) = p_n(x_n) - p_{n0}$ and $\Delta n_p(-x_p) = n_p(-x_p) - n_{p0}$. From figure 3.14b we can see that the bottom of the conduction band E_C in the N and P sides in equilibrium differ in energy by eV_{bi}. Hence, assuming that Boltzmann statistics can be used instead of Fermi–Dirac statistics, we can write for the concentration of electrons at the conduction band minimum at the respective sides,

$$n_{n0} = n_{p0}exp(eV_{bi}KT)$$ 3.51a

and similarly

$$p_{p0} = p_{n0}exp(eV_{bi}KT)$$ 3.51b

where we have used the subscript 0 to denote equilibrium values. Rearranging 3.51a we have

$$V_{bi} = \frac{KT}{e}ln\left(\frac{n_{n0}}{n_{p0}}\right)$$

Using the relation $n_{no}p_{no} = n_i^2$ that we have proved in chapter 2 (equation 2.61), we get

$$V_{bi} = \frac{KT}{e}ln\left(\frac{n_{n0}p_{p0}}{n_i^2}\right) = \frac{KT}{e}ln\left(\frac{N_A N_D}{n_i^2}\right)$$ 3.52

We have therefore obtained the built-in voltage V_{bi} in terms of the doping levels N_A, N_D and the equilibrium concentration n_i.

When the voltage V is applied to the junction and the barrier reduces to $V_{bi} - V$, we can follow exactly the same argument as above and we then for electrons facing the reduced barrier get

$$n_n(x_n) = n_p(-x_p)exp\left[e(V_{bi} - V)/KT\right]$$

$$\Rightarrow n_p(-x_p) = n_n(x_n)exp(-eV_{bi}kT)exp(eV/KT)$$

But as we have repeatedly noted, the majority carrier concentration n_n is hardly changed by the injection and $n_{n0} = n_n$. So using 3.51 we get

$$n_p(-x_p) = n_{p0}exp(eV/KT)$$ 3.53a

or equivalently

$$\Delta n_p(-x_p) = n_{p0}\left[exp(eV/KT) - 1\right]$$ 3.53b

This constitutes the exponential increase we have referred to in the previous paragraph. Similarly, we also get

$$\Delta p_n(x_n) = p_{no}\left[exp(eV / KT) - 1\right] \qquad 3.53c$$

assuming likewise $p_p = p_{po}$.

Therefore, we have obtained the preexponential factor in the injected hole carrier concentration, equation 3.50, and we get

$$\Delta p_n(x) = \Delta p_n(x_n) exp\left[\frac{-(x - x_n)}{L_p}\right] = p_{no}\left[exp(eV / KT) - 1\right] exp\left[\frac{-(x - x_n)}{L_p}\right] \qquad 3.54a$$

and similarly for

$$\Delta n_p(x) = \Delta n_p(-x_p) exp\left[\frac{(x + x_p)}{L_n}\right] \qquad 3.54b$$

We therefore have

$$J_p^{diff}(x) = -eD_p\frac{dp}{dx} = \frac{eD_p p_n(x_n)}{L_p} exp\left[\frac{-(x - x_n)}{L_p}\right] \qquad 3.55a$$

and

$$J_n^{diff}(x) = eD_n\frac{dn}{dx} = \frac{eD_n n_p(-x_p)}{L_n} exp\left(\frac{x + x_p}{L_n}\right) \qquad 3.55b$$

The injected current density into the neutral P region and into the neutral N region can now be evaluated by taking the values of the above at $x = -x_p$ and $x = x_n$ respectively. We get using 3.53a and 3.53b

$$J_p(x_n) = \frac{eD_p p_{no}}{L_p}\left[exp(eV / KT) - 1\right] \qquad 3.56a$$

and

$$J_n(-x_p) = \frac{eD_n n_{po}}{L_n}\left[exp(eV / KT) - 1\right] \qquad 3.56b$$

The total current will be equal to the sum of the above two components. We will have

$$J = J_p(x_n) + J_n(-x_p) = J_0\left[exp\left(\frac{eV}{KT}\right) - 1\right] \qquad 3.57$$

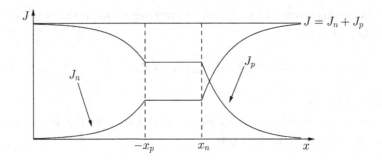

FIGURE 3.17   Electron, hole, and total current densities across an ideal PN junction.

where

$$J_0 = \frac{eD_p p_{n0}}{L_p} + \frac{eD_n n_{p0}}{L_n}$$ 3.58

The variation of $J_p(x)$ and $J_n(x)$ in the entire range of x is shown in figure 3.17, while the total current density as a function of voltage is shown in figure 3.18. An examination of figure 3.17 shows that apart from the minority carrier currents, there are also majority carrier currents, as indicated by the presence of J_n in the N-type region and J_p in the P-type region. Where do these currents come from? The total current density J must be constant and independent of x and these currents actually guarantee the constancy of J over the distance x. Equations 3.55a and 3.55b are valid only when they describe minority carrier movement, i.e. in the regions $x > x_n$ and $x < -x_p$ respectively (when they describe an exponential decrease with x). When minority carriers are injected, let us say a concentration $\Delta p(x)$ of holes in the N-side, an equal amount of electrons $\Delta n(x)$ is generated to preserve the neutrality of the region and this $\Delta n(x)$ in the N region creates an extra current $J_n(x)$ of majority carriers. However, this current does not obey equations like 3.55a and 3.55b because these equations were derived from the continuity equations in the specific form valid only for

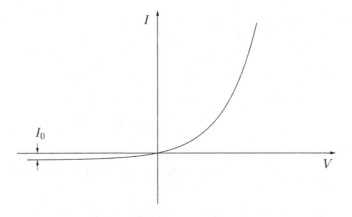

FIGURE 3.18   I-V characteristic of an ideal PN junction.

minority carriers. Actually, the majority carrier currents contain a significant amount of drift because our initial assumption of the applied voltage dropping exclusively in the depletion layer was only an approximation and a small electric field exists in the neutral regions. However, by adding the diffusion currents at the respective edges of the depletion layer we have taken care of all currents.

3.9 THE NON-IDEAL, REAL PN JUNCTION

One of the assumptions we have made in deriving the ideal PN junction current equation is that there is no recombination in the depletion layer when the majority carriers are injected into the neutral P and N regions. This is not actually true. The semiconducting materials always suffer from native metal impurities or even defects produced during the doping process which give rise to mid-gap states which in turn produce generation and recombination processes in the depletion region. These in turn give extra currents that we have not taken into account so far.

We have discussed recombination through mid-gap states in section 3.4. Equation 3.24 gives this rate as being equal to

$$R_t = \frac{np}{\tau_n n + \tau_p p} \quad \text{3.24 reproduced here}$$

The maximum recombination rate will develop when the denominator is a minimum. This in turn implies that $n \approx p$. Let us now look at the value of np at the two ends of the depletion layer. We have from equations 3.53a-c

$$n_p\left(-x_p\right) = n_{po} e^{eV/KT}$$

and

$$p_p\left(-x_p\right) = p_{po}$$

Therefore

$$n_p\left(-x_p\right) p_p\left(-x_p\right) = n_i^2 exp\left(\frac{eV}{KT}\right) \quad\quad\quad 3.59a$$

and by a similar argument

$$p_n\left(x_n\right) n_n\left(x_n\right) = n_i^2 exp\left(\frac{eV}{KT}\right) \quad\quad\quad 3.59b$$

Although the equality of the product pn has been established only at the end points of the depletion layer, numerical calculations do in fact show that the product pn remains constant in the depletion region.

We then obtain assuming that approximately $\tau_n = \tau_p = C_r$

$$R_t = \frac{n}{2C_r} = \frac{n_i}{2C_r} exp\left(\frac{eV}{2KT}\right)$$

3.60

The recombination R_t above denotes processes per unit time per unit volume so that an integration over the volume of the depletion layer is required to get the recombination current I^R. Hence if A denotes the cross-section of the diode and W, as usual, the length of the depletion region

$$I^R = \frac{eAW}{2C_r} exp\left(\frac{eV}{2KT}\right) = I_0^R exp\left(\frac{eV}{2KT}\right)$$

3.61

In addition to the recombination current there is also a generation current I_0^G and the observed current is the difference between the two. We note that the generation current I_0^G is depends only on temperature. At $V = 0$ the two must balance. Therefore

$$I^{GR} = I^R - I^G = I_0^R\left[exp\left(\frac{eV}{2KT}\right) - 1\right]$$

3.62

This current must be added to the diffusion current we calculated previously, so we get

$$I = I_0\left[exp\left(\frac{eV}{KT}\right) - 1\right] + I_0^R\left[exp\left(\frac{eV}{2KT}\right) - 1\right]$$

3.63

The sum of the above two currents in equation 3.63 can be written more compactly in an empirical form

$$I = \widetilde{I_0^R}\left[exp\left(\frac{eV}{mkT}\right) - 1\right]$$

3.64

where $1 \leqslant m \leqslant 2$. At low values of the applied voltage V, the current I is dominated by the recombination term and $m \approx 2$ whereas at higher V the current I is essentially composed of the diffusion current and $m \approx 1$. The constant m is usually called the "ideality factor" of the diode. A schematic graph of the total current I versus the voltage V is shown in figure 3.19.

From figure 3.19 it can be seen that at very high voltages the rate of increase of I falls below the rate dictated by $m = 1$. This is due to the fact that at such voltages, all of the applied V does not drop nearly exclusively in the space charge/depletion layer and there is some part of V – let's call it $\tilde{V}$ – which drops inside the neutral P and N regions. Then

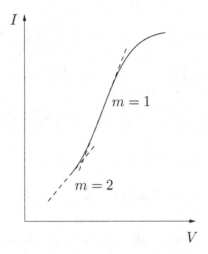

FIGURE 3.19   I-V characteristic of a real PN junction.

the barrier inside the depletion region is not $\left[V_{bi} - V\right]$ but it is equal to $\left[V_{bi} - \left(V - \tilde{V}\right)\right]$. This would lead to the diffusion current being equal to

$$I \approx I_0 exp\left[\frac{e\left(V - \tilde{V}\right)}{KT}\right] = \frac{I_0\, exp\left(\dfrac{eV}{KT}\right)}{exp\left(\dfrac{e\tilde{V}}{KT}\right)} \qquad 3.65$$

and the current increases at a slower rate as equation 3.65 indicates. There are also temperature effects in PN junctions as both diffusion and generation–recombination are temperature dependent processes which are well documented having been explored since the 1950s. We will not deal with these as the purpose of this book is to explore the transition from the mechanisms of conduction which treat the electrons as particles to those that need to treat electrons as waves, as is necessary in modern devices.

3.10 THE METAL–SEMICONDUCTOR OR SCHOTTKY JUNCTION

The metal–semiconductor junction is a diode just like the PN junction, i.e. it carries a very low current in reverse bias and a very large current in forward bias. It also carries many of the physical characteristics of the PN junction and in a much simpler way. It is sometimes called a Schottky diode after the German physicist Schottky who first analyzed it before the second world war. Figure 3.20 shows the simplified band diagrams of a metal piece and an N-type semiconducting piece just before touching.

An important quantity that has been added to these simplified band diagrams of the metal and the semiconductor, compared to what has been presented so far, is the work-function. If periodic boundary conditions are imposed on a crystalline solid, its eigen-values $E_l(\mathbf{k})$ extend from zero energy to infinity with the band index l having no bound. However real solids do not obey periodic boundary conditions—in 3-dimensions it is

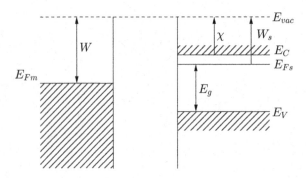

FIGURE 3.20   Band structure of a metal and an N type semiconductor before contact.

impossible to deform a solid to make all its faces meet—and solids are terminated at surfaces. A consequence of this is that there is a maximum of energy the electrons can have to
remain in the solid. If more energy is gained, by photoexcitation for example, the electrons
get out of the solid.

This maximum energy is usually called the vacuum energy and denoted E_{vac} as shown
in figure 3.20. The difference $W = E_{vac} - E_{F_m}$, where E_{F_m} is Fermi level of the metal, is the
workfunction of a metal and it corresponds to the minimum energy required to extract
an electron from the metal. The case of a semiconductor is somewhat different since
the Fermi-level E_{F_s} lies usually in the energy-gap. Then a second quantity needs to be
defined, called electron affinity, which is the difference $\chi = E_{vac} - E_C$. The workfunction
$W_s = E_{vac} - E_{F_s}$ is still defined for a semiconductor, but we normally work with the affinity
χ when it comes to semiconductor interfaces. All these quantities of a semiconductor are
shown in figure 3.20.

Returning now to the problem of the junction, we observe that the N-type semiconductor
Fermi level is above the Fermi level of the metal, so when the two materials come into
contact electrons will flow from the semiconductor to the metal so as to equalize the two
originally different Fermi levels as happened in the case of the PN junction. This will create
a deficiency of electrons in the N-type semiconductor and a surplus in the metal, i.e. a
space charge layer will develop with a built-in potential inside, again as in the PN junction.
Only this time, the space charge layer will develop mainly in the semiconductor for the
following reason: the metal has a much higher density of states than the semiconductor,
so that it will take a very small volume of it to accommodate the electrons that have flown
from the semiconductor. This is shown in figure 3.21a, where we plot the charge density ρ
across the junction. Again, as in the PN junction, we have approximated the exact charge
density $\rho = e\left(N_d^+ - n\right)$ with $\rho = N_D^+$.

As usual the electrostatics of the junction will be governed by Poisson's equation. We
choose to ignore the very small negative surface layer on the metal side so that Poisson's
equation reads on the semiconductor side

$$\frac{d^2V}{dx^2} = \frac{-eN_D^+}{\varepsilon} = \frac{-eN_D}{\varepsilon} \quad 0 \leqslant x \leqslant d \qquad 3.66$$

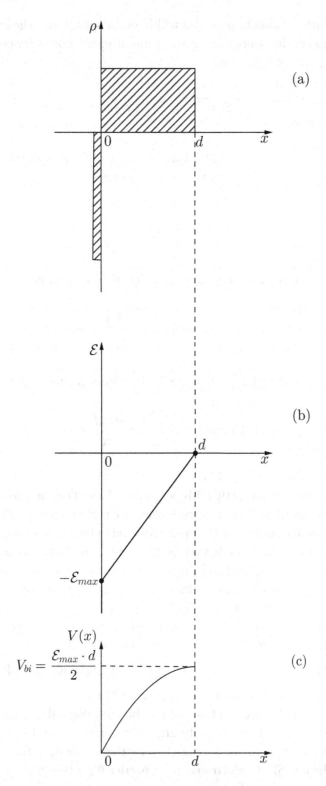

FIGURE 3.21 Charge density (a), electric field (b), and potential (c) across a metal-semiconductor (Schottky) junction.

where d is the length (or sometimes called width) of the space charge layer and we assume complete ionization of the donors. Integrating and using the boundary condition that the electric field $\mathcal{E}$ is zero at $x = d$ we get

$$\mathcal{E} = \frac{-dV}{dx} = \frac{eN_D}{\varepsilon}(x - d) \qquad 3.67$$

The plot of $\mathcal{E}(x)$ is shown in figure 3.21b. The maximum of the absolute value of the electric field occurs at $x = 0$. If we call this maximum $\mathcal{E}_{max}$ we get

$$\mathcal{E}(x) = \mathcal{E}_{max}\left(\frac{x}{d} - 1\right) \qquad 3.68$$

Integrating $\mathcal{E}(x)$ and using as a boundary condition $V(0^+) = 0$ we get

$$V(x) = \mathcal{E}_{max}\left(x - \frac{x^2}{2d}\right) \qquad 3.69$$

From either the area of the triangle in figure 3.21b or from equation 3.69 above

$$V(d) \equiv V_{bi} = \frac{\mathcal{E}_{max}d}{2} = \frac{eN_D d^2}{2\varepsilon} \qquad 3.70$$

The plot of $V(x)$ is shown in figure 3.21c.

To construct the band bending that this potential will exert on the semiconductor bands we simply have to multiply by $(-e)$. Remember electrostatic energy = charge × potential. Furthermore note that (see figure 3.20) the Fermi level in the metal differs by the constant energy difference $W - \chi$. from the bottom of the conduction band in the semiconductor. Therefore, the electronic energy band diagram of the metal–semiconductor junction is shown in figure 3.22a. It is left to the reader as an exercise to prove that the potential in the N-type semiconductor of the Schottky diode is the same as the potential in the N-side of a PN junction when overdoped in the P-side (usually denoted P^+N). The band diagram of a metal P-type semiconductor follows the same principles and equations, only the RHS of the Poisson equation is equal to $\left(+\frac{eN_A}{\varepsilon}\right)$ so that finally the bands in the P side are upward bending instead of downward as is shown in figure 3.22b.

Let us now examine the electron flow with no bias on. We will restrict ourselves to the metal N-type semiconductor junction. The arguments for the metal P-type semiconductor are exactly the same. At zero bias there are two barriers to electron flow. From the metal (M) to semiconductor (S), the electrons face a barrier $\Phi_B = W - \chi$. From S to M, on the other hand, the electrons face the barrier V_{bi}, see figure 3.22a. The barrier Φ_B is called the Schottky barrier of the diode. The current must be zero at zero bias. This current is indeed

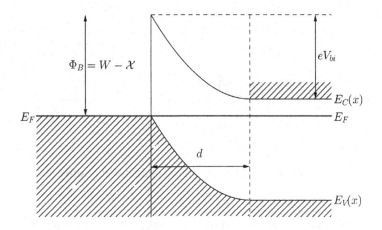

FIGURE 3.22a   Band structure of a metal N-type semiconductor junction at zero bias.

the sum of the currents $I_{M \to S}$ and $I_{S \to M}$ in an obvious notation, each one corresponding to one of the electron flows described above.

We will assume that the relation for the electron density

$$n = N_C exp\left(\frac{E_F - E_C}{KT}\right)$$

which holds for a homogeneous semiconductor, also holds for every point in an inhomogeneous one. Looking at figure 3.22a, we see that the electron density at the surface is smaller by $exp(-eV_{bi} / kT)$, compared to the bulk, because $(E_C - E_F)$ is bigger by V_{bi} at the surface compared to the bulk. Therefore the electron density at the interface n_{if} is

$$n_{if} = N_D exp\left(\frac{-eV_{bi}}{KT}\right) \qquad\qquad 3.71$$

This electron density will generate a current $I_{S \to M}$ from $S \to M$. An equal in magnitude current $I_{M \to S}$ will flow of course in the opposite direction.

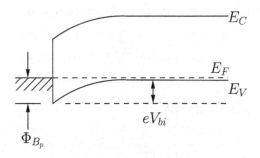

FIGURE 3.22b   Band structure of a metal P-type semiconductor.

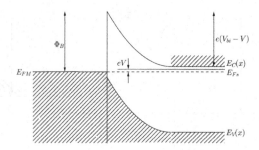

FIGURE 3.22c Band structure of a metal N-type semiconductor at positive bias.

Now if a positive voltage is applied to the junction (positive meaning the + electrode of the generator connected to the metal side), the barrier (eV_{bi}) will be reduced to $e(V_{bi} - V)$. The same argument we gave for the PN junction holds here as well, see figure 3.22c. In this figure E_{Fm} and E_{Fs} stand for the Fermi levels of the metal and semiconductor respectively. On the other hand the barrier Φ_B from $M \rightarrow S$ will remain the same. We have obviously assumed that the applied voltage V drops to a very good approximation entirely in the semiconductor part of the junction. As a result of the previous arguments, the current $I_{M \rightarrow S}$ will remain the same while $I_{S \rightarrow M}$ will increase by $exp(eV/kT)$. The total current therefore will be

$$I(V) = -I_{M \rightarrow S} + I_{S \rightarrow M} exp\left(\frac{eV}{kT}\right) = I_0\left[exp\left(\frac{eV}{kT}\right) - 1\right] \qquad 3.72$$

where $I_0 = |I_{M-S}|$. The graph of $I(V)$ for the Schottky diode is the same as that of a PN junction, see figure 3.18.

Both the PN junctions and M–S Schottky junctions are used as rectifiers since the current in the negative voltage domain is negligible. The M–S junctions, however, are everywhere in microelectronic circuits as metal contacts because current must be thrown in and taken out of devices. This happens through metal contacts. Most of them have to be ohmic while the M–S junction is rectifying. Therefore, it is necessary to be able to transform the rectifying contact into an ohmic one. Indeed, overdoping of the semiconductor part of the M–S junction will produce an ohmic contact. How can this be done?

From equation 3.70 we have that if the applied voltage is zero

$$d = \sqrt{\frac{2\varepsilon V_{bi}}{eN_D}} \qquad 3.73$$

and if a voltage is applied to the Schottky diode

$$d = \sqrt{\frac{2\varepsilon(V_{bi} - V)}{eN_D}} \qquad 3.73a$$

The built in voltage V_{bi} is practically constant with doping level N_D. To see this, we observe from figure 3.22a or 3.22c that

$$\Phi_B = eV_{bi} + E_C - E_F \qquad 3.74$$

The barrier Φ_B is a material property of the interface equal to $(0.8 - 0.9)eV$ whereas $E_C - E_F \approx 40 - 50meV$, so V_{bi} is practically constant. Hence overdoping the semiconductor by two more orders of magnitude than normal will reduce the barrier width d by a factor of 10. If the barrier width is of the order of $2 - 3nm$, tunneling from the metal to the semiconductor can take place, see figure 3.23.

The tunneling probability can be calculated by the general WKB formula, equation 1.65 derived in chapter 1, with the vacuum mass m replaced by the effective mass m^*. If an electron stays within the semiconductor, it experiences a potential given by equation 3.69. But if it crosses the metal–semiconductor interface from the metal side, it experiences the barrier Φ_B, therefore the tunneling potential energy U(x) experienced by an electron tunneling from the metal into the semiconductor is

$$U_{tun}(x) = \Phi_B - eV(x) =$$

$$= \Phi_B - e\mathcal{E}_{max}\left(x - \frac{x^2}{2d}\right) \qquad 3.75$$

and the probability of tunneling from the Fermi level E_{Fm}, see figure 3.22, is

$$T = exp\left[\frac{-2}{\hbar}\int_{x_1}^{x_2}\sqrt{2m^*\left(U_{tun}(x) - E_{Fm}\right)}dx\right] \qquad 3.76$$

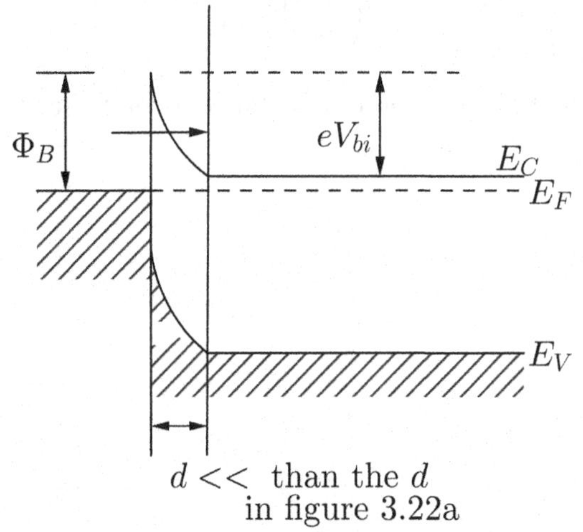

$d \ll$ than the d
in figure 3.22a

FIGURE 3.23   Same as figure 3.22a but with the semiconductor layer being overdopped.

The critical reader may note that the applied voltage V does not appear explicitly in equations 3.75, 3.76. It does so indirectly through the quantity $\mathcal{E}_{max}$. From figure 3.21b we get $\mathcal{E}_{max}d/2 = V_{bi} - V$ when an external voltage V is on, so the latter affects the transmission probability T.

PROBLEMS

3.1 The mobility of a pure sample of GaAs is measured and found to be μ_1. The sample is then doped and its mobility measured again and found to be μ_2. Assume that the doping does not change its effective mass m*. Find an expression for the relaxation time of the electrons due to collisions to the dopant atoms.

3.2 Prove that the hole current density

$$J_p = J_p^{drift} + J_p^{diff}$$

may be written compactly in one dimension as

$$J_p = p\mu_p \frac{d(E_{Fp}(x))}{dx}$$

3.3 Find the built-in potential V_{bi} and width w of the depletion region of a linearly doped PN function, that is of a PN function whose doping is $N(x) = cx$ where c is a constant and x lies $-\frac{w}{2} < x < \frac{w}{2}$.

3.4 By imposing the conditions that at equilibrium ($V = 0$) the currents J_n, J_p are each separately zero obtain the expressions for the built-in potential

$$V_{bi} = \frac{KT}{e} \ln\left(\frac{n_{n0}}{n_{p0}}\right) = \frac{KT}{e} \ln\left(\frac{p_{p0}}{p_{n0}}\right)$$

3.5 Show that the potential variation of a metal-semiconductor junction is the same as that of an overdoped P⁺N junction.

CHAPTER 4

Advanced Classical Theory of Conduction

4.1 THE NEED FOR A BETTER CLASSICAL THEORY OF CONDUCTION

In section 2.6 we proved that under the action of an electric field $\mathcal{E}$, an electron accelerates as if it were a classical particle, i.e. its acceleration given by $e\mathcal{E}$ divided by the effective mass m^* in 1 dimension and in 3 dimensions by $e\mathcal{E}$ with each component divided by the corresponding diagonal element of the effective mass tensor m_{ij}^* (see equation 2.49b). In chapter 3, on the other hand, we have seen that the motion of an electron under the action of $\mathcal{E}$ is not that of perpetual acceleration and because of the presence of collisions a steady velocity appears for the ensemble of electrons that we call drift velocity. We gave an elementary proof of that in section 3.2. The proof ignored the existence of k space and Bloch functions since the treatment there was intentionally elementary. A qualitative picture of the application of $\mathcal{E}$ and collisions in terms of k space was described in figures 3.2a and 3.2b, which portray a situation where more electrons travel against $\mathcal{E}$ with higher velocities than in the opposite direction. However that was not proven. It is time now to acquire a deeper understanding by studying the Boltzmann equation which allows electric fields, diffusion, and collisions to be examined simultaneously in a rigorous manner within the framework of k space.

We saw in section 2.5 that when an electric field $\mathcal{E}$ is applied to a crystalline solid the wavefunctions are no longer Bloch waves, which fill homogeneously the whole solid, but wavepackets, i.e. localized in space states, see equation 2.37. Hence the notion of acceleration and collision for particles has meaning because it is impossible to reconcile the notion of collision at some point in space with a wavefunction for electrons that extends from one end of the crystal to the other. We now rewrite equation 2.37 more analytically in terms of the components of a Bloch wave.

$$\Psi(r,t)=\int A(k)u(k,r)e^{i\left[k\cdot r-\frac{E(k)t}{\hbar}\right]}d^3k \qquad 4.1$$

where $d^3k = dk_x dk_y dk_z$ and $A(k)$ is sharply peaked around a given $k = k'$ with $\hbar k'$ being the momentum of that wavepacket if the bands are parabolic. The function $u(k,r)$ is the periodic part of the Bloch function. As we have seen in section 2.5, such a wavepacket is also sharply localized in space around a maximum point $r = r_{max}$. Furthermore, it is easy to prove (see problem 4.1) that the group velocity of this particular wavepacketis in an arbitrary band is

$$v(k) = \frac{1}{\hbar} \nabla_k \big(E(k) \big) \qquad\qquad 4.2$$

where by the symbol ∇_k we mean the gradient in k space. From the properties of wavepackets presented above, we conclude that for the rest of this chapter we can forget the wave properties of electrons and assume that they can be treated as classical particles with their position given by the point of maximum in 4.1 (or the point of average value) and their velocity by 4.2 above. The validity of this approximation is examined below.

4.2 THE BOLTZMANN EQUATION

When an equilibrium is established the probability of finding an electron with energy E depends only on E and is given by the Fermi–Dirac distribution. This may seem like a tautology but what it means is that if we have two states with two distinct wavevectors $k_1 \neq k_2$ with $E(k_1) = E(k_2)$, the probability of occupation of the states is the same, that is, it does not depend on the wavevector or crystal momentum. This is not the case when non-equilibrium processes are present. In fact, the Boltzmann equation is the equation that governs the probability $f(k,r,t)$ at non-equilibrium of finding an electron at time t with wavevector k inside an infinitesimal volume around r. This classical description is founded quantum mechanically on equations 4.1 and 4.2 of the previous section. For electrons to occupy such simple Bloch wavepackets, the system under investigation must be macroscopically large (of the order of fractions of microns) in each dimension. If one or more dimensions reach the nanometer scale, then additional complications arise which we will deal with in the next section. The requirement that the system be large is not the only one which prohibits the use of the Boltzmann equation. A more demanding requirement is that any applied fields or potentials must vary slowly—the wavelength of any applied electric field $\mathcal{E}(r)$ must be large compared to the width of the wavepackets representing the electrons. Figure 4.1 gives a pictorial representation of this statement. The necessity of this restriction comes from the fact that the notion of point-like particles acted on by the local field at point r is no longer valid. Then one has to abandon the Boltzmann equation altogether as we will do in the next chapter.

The probability or distribution function $f(k,r,t)$ can change in 3 different ways: a) an external electric field can change the wavevector k of any electron cf. equation 2.41, b) an

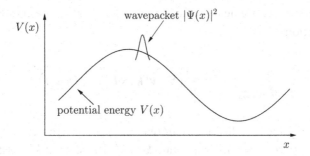

FIGURE 4.1   Requirements for a particle picture to be valid: the extent of a wavepacket representing a particle must be much smaller than the wavelength of the potential energy distribution.

initially non-uniform distribution may produce diffusion and change r in $f(k,r,t)$, and c) collisions abruptly change the wavevectors of the electrons. Therefore, we can write

$$\frac{df}{dt} = \frac{\partial f}{\partial t}\bigg|_{externalfield} + \frac{\partial f}{\partial t}\bigg|_{diffusion} + \frac{\partial f}{\partial t}\bigg|_{collisions} \qquad 4.3$$

Before we continue, a small point on notation. As we will need both the gradient in real space and in k space, we will differentiate between them by denoting them as ∇_r and ∇_k respectively. We now assume that the distribution of carriers in the neighbourhood of r at time t is equal to the distribution in the neighbourhood of $r - vt$ at time 0. Then

$$f(k,r,t) = f(k,r-v(k)t,0) \qquad 4.4$$

Then the second term in equation 4.3 becomes

$$\frac{\partial f}{\partial t}\bigg|_{diffusion} = -v(k)\nabla_r f(k,r,t) \qquad 4.5$$

In a similar manner to equation 4.4 we have

$$f(k,r,t) = f(k-\dot{k}t,r,0) \qquad 4.6$$

where $\dot{k}$ denotes the time derivative of k. We then get for the first term in 4.3

$$\frac{\partial f}{\partial t}\bigg|_{externalfield} = -\dot{k}\nabla_k f(k,r,t) = \frac{-F}{\hbar}\nabla_k f(k,r,t) = \frac{-q\mathcal{E}}{\hbar}\nabla_k f(k,r,t) \qquad 4.7$$

where, in equation 4.7 we have used equation 2.41 indicating the rate of change of the crystal momentum $\hbar k$ is equal to the electric force exerted on the crystal. Note that the charge q may be either negative or positive so far. We have written $F = q\mathcal{E}$ above but in fact the magnetic as well as the electric field produce a rate of change of the crystal momentum. So we generalize equation 4.7 to

$$\frac{\partial f}{\partial t}\bigg|_{external field} = \frac{-q}{\hbar}\left(\mathcal{E}+v(k)\times B\right)\nabla_k f(k,r,t)$$

4.8

The rate of change of f due to collisions is more complicated than the first two terms in 4.3. A collision, say, transfers electrons from state k to k'. The state at k' must be empty for an electron to fall in and, likewise, the state at k must be occupied. Therefore, the rate of transitions depends on the occupation probabilities of the individual states. Furthermore, there is also a reverse transition from k' to k and all possible pairs (k,k') must be considered to account for the rate of change of f due to collisions. Using for the moment the abbreviation

$$f_k = f(k,r,t)$$

we have

$$\frac{\partial f}{\partial t}\bigg|_{collision} = \int P_{k,k'} f_{k'}\left(1-f_k\right) - P_{k',k} f_k\left(1-f_{k'}\right)d^3k'$$

4.9

where in 4.9 $P_{k,k'}$ is the probability of transition from k' to k if k' is known to be completely occupied and k completely empty. $P_{k,k'}$ and $P_{k',k}$ are equal in equilibrium but they are also equal if the transitions from k' to k and vice versa are elastic. We limit ourselves for the moment to elastic collisions. Putting expression 4.5, 4.8, and 4.9 in equation 4.3 will result in the time dependent Boltzmann equation, an integrodifferential equation which can be solved in a limited number of cases that are, however, of paramount importance. The Boltzmann equation then reads after substituting for $v(k) = \frac{1}{\hbar}\nabla_k E(k)$

$$-\frac{df}{dt} = q\left(\mathcal{E}+\frac{1}{\hbar}\nabla_k E(k)\times B\right)\frac{1}{\hbar}\nabla_k f(k,r) + \frac{1}{\hbar}\nabla_k E(k)\nabla_r f(k,r) -$$
$$-\int P_{k,k'}\left[f(k',r)-f(k,r)\right]dk'$$

4.10

Note that in equation 4.10 we have omitted the products $f_k * f_{k'}$ in 4.9, which cancel out.

4.3 SOLUTION OF THE BOLTZMANN EQUATION BY THE RELAXATION TIME APPROXIMATION

As noted previously, equation 4.10 is an integro-differential equation and therefore some drastic approximations have to be made to obtain an analytical solution. However, the labour is worthwhile not only because of the greater numerical accuracy compared to empirical models but mainly because of the greater physical insight obtained by the analytical solutions.

We are mostly interested in solutions where the derivative df/dt is zero. Note that these are steady-state solutions, not equilibrium solutions, energy is being driven into the system by the fields. The most difficult term is obviously the integral term of the RHS of 4.10. The so-called relaxation time approximation simplifies this last term by putting

$$\left.\frac{\partial f}{\partial t}\right|_{collisions} = -\frac{f-f_0}{\tau}$$ 4.11

i.e. the rate of change due to collisions is proportional to the deviation from the equilibrium distribution f_0—the Fermi–Dirac distribution

$$f_0(E) = \frac{1}{1+exp\left(\dfrac{E-E_F}{KT}\right)}$$

What is the meaning of equation 4.11 and when is it valid?

Let us write for a homogeneous system

$$f = f_0 + f'$$ 4.12

where f' is the deviation from the equilibrium value f_0 and assume that the system is perturbed by some field which is turned off at some time later. Then the first two terms of the RHS of 4.10 are zero and we have

$$\frac{df}{dt} = \frac{df'}{dt} = -\frac{f'}{\tau}$$ 4.13

$$f'(t) = f'(0)e^{-t/\tau}$$

i.e. the system returns back to its equilibrium distribution exponentially with time, with τ being the approximate time constant. So the relaxation approximation looks like a reasonable assumption. But when is it valid?

For the relaxation approximation to be valid, the deviation from equilibrium must be small, which in turn means that the fields acting on the system must be small. Since the latter statement is not always quantifiable, it is a safer condition that the relaxation time τ, which in principle can be a function of k, must be independent of the strength of the perturbation causing the departure from equilibrium. A rather long argument shows that if only elastic processes are involved then the use of the relaxation time $\tau(k)$ is a valid approximation. The use of the relaxation time approximation for inelastic processes is not valid. We will not delve into this as we are going to abandon the Boltzmann equation in the next chapter. But given the limitations described above, we are now ready to obtain solutions of the

Boltzmann equation that shed more physical insight in what we have done so far and analyze situations where the previous semiempirical approach was insufficient.

Let us first rewrite for the collision term

$$\frac{\partial f}{\partial t}\bigg|_{collision} = -\frac{f'}{\tau} \qquad\qquad 4.14$$

and for the deviation from equilibrium f'

$$f' = -\Phi \frac{\partial f_0}{\partial E} \qquad\qquad 4.15$$

where Φ can be a function of k and r. Using the definition 4.14, equation 4.10 becomes in the steady state $df/dt = 0$, with the velocity restored back in the equation.

$$q(\mathcal{E}+v\times B)\cdot\frac{\nabla_k f}{\hbar}+v\cdot\nabla_r f = \frac{\Phi}{\tau}\frac{\partial f_0}{\partial E} \qquad\qquad 4.16$$

Now a series of approximations are necessary to solve 4.16. We first assume that the perturbing fields do not induce any space dependent variation in f'. There can be a space dependent variation in the total f but only if there is originally one due to a thermal gradient. In this case, both the temperature $T(r)$ and the Fermi level $E_F(r)$ will be functions of r. Then $f_0(r)$ will be a function of r. The use of the Fermi–Dirac distribution, with $T(r)$ and $E_F(r)$ being space dependent, is only an approximation that is valid if the variations in $T(r)$ and $E_F(r)$ are so weak that the region around any r may be considered as being in pseudo equilibrium with its neighbouring regions. In mathematical terms

$$\nabla_r f \approx \nabla_r f_0 = \frac{\partial f_0}{\partial\left(\dfrac{E}{KT}\right)}\nabla_r\left(\frac{E-E_F(r)}{KT(r)}\right) \qquad\qquad 4.17$$

where K = Boltzmann's constant.

The reason for writing equation 4.15 in this product form should now be clear. We expect $\partial f_0/\partial E$ to appear as a common term in all our manipulations. From equation 4.17 after differentiating the quotient we get

$$\nabla_r f = -\left(\left(\frac{E-E_F}{T}\right)\cdot\nabla_r T+\nabla_r E_F\right)\frac{\partial f_0}{\partial E} \qquad\qquad 4.18a$$

or

$$\nabla_r f = -\left[\frac{E}{T}\cdot\nabla_r T+T\nabla_r\left(\frac{E_F}{T}\right)\right]\frac{\partial f_0}{\partial E} \qquad\qquad 4.18b$$

For the term $\nabla_k f$ we have,

$$\nabla_k f = \frac{\partial f_0}{\partial E}\nabla_k E(k) + \nabla_k f' = \frac{\partial f_0}{\partial E}\hbar v(k) + \nabla_k f' \qquad 4.19$$

Substituting equations 4.18a, and 4.19 into 4.16 we get

$$q\mathcal{E}\cdot v\frac{\partial f_0}{\partial E} + q(v\times B)\cdot v\frac{\partial f_0}{\partial E} + \frac{q}{\hbar}(v\times B)\cdot\nabla_k f' - \left[(E-E_F)\nabla_r ln(T) + \nabla_r E_F\right]\cdot v\frac{\partial f_0}{\partial E} = \frac{\Phi}{\tau}\frac{\partial f_0}{\partial E} \qquad 4.20$$

Note that in 4.20 we have dropped the term multiplying the electric field $\mathcal{E}$ with $\nabla_k f'$ since it is a term multiplying the perturbing field with the perturbation and hence it must be of second order to the first term involving the electric field. Note also that the second term in 4.20 is zero because it involves the inner product of two perpendicular vectors. Rearranging 4.20 we get

$$\left(q\mathcal{E} - \left[(E-E_F)\nabla_r ln(T) + \nabla_r E_F\right]\right)\cdot v\frac{\partial f_0}{\partial E} + \frac{q}{\hbar}(v\times B)\nabla_k f' = \frac{\Phi}{\tau}\frac{\partial f_0}{\partial E} \qquad 4.21$$

We now need to express the second term on the LHS of 4.21 in a product form with $\frac{\partial f_0}{\partial E}$ so that the latter can be canceled from both sides of 4.21. The following manipulations are elementary

$$\frac{q}{\hbar}(v\times B)\cdot\nabla_k f' = \frac{q}{\hbar}B\cdot\left(\nabla_k f'\times v\right) = \frac{-q}{\hbar^2}B\cdot\left(\nabla_k E\times\nabla_k f'\right) = \frac{q}{\hbar^2}B\cdot\left(\nabla_k E\times\nabla_k\Phi\right)\frac{\partial f_0}{\partial E} \qquad 4.22$$

where in the last step of 4.22 we have used the definition $f' = -\Phi\partial f_0/\partial E$. Combining now 4.21 and 4.22 we get

$$\Phi(k,r) = \tau(k)\Big(q\mathcal{E} - \left[(E-E_F)\nabla ln(T) + \nabla_r E_F\right]\Big)\cdot v(k) +$$

$$+ \frac{q\tau(k)}{\hbar^2}B\cdot\left(\nabla_k E(k)\times\nabla_k\Phi(k,r)\right) \qquad 4.23$$

where in 4.23 we have reinstituted the k and r dependence where appropriate. Equation 4.23 is the linearized Boltzmann equation that will form the basis of our subsequent investigations.

4.4 APPLICATION OF AN ELECTRIC FIELD–CONDUCTIVITY OF SOLIDS

When only an electric field is applied to a solid with no temperature gradient and no magnetic field, we have $\nabla T = 0$, $B = 0$ and $\nabla E_F = 0$. At first sight the equality $\nabla E_F = 0$ may seem odd since if there is an electric field there must be a potential difference and a corresponding ΔE_F.

But the essential approximation we are making is that the change in the non-equilibrium distribution function $f(\mathbf{k}, \mathbf{r})$ has no $\mathbf{r}$ dependence and most of the change $f' = f - f_0$ comes from the $\mathbf{k}$ – dependence which derives solely from the electric field $\boldsymbol{\mathcal{E}}$. Under these circumstances we can neglect any variation in the Fermi level. We then get straight away from equation 4.23

$$\Phi = q\tau\boldsymbol{\mathcal{E}} \cdot \mathbf{v} \qquad 4.24$$

and

$$f = f_0 - q\tau\boldsymbol{\mathcal{E}} \cdot \mathbf{v}\frac{\partial f_0}{\partial E} \qquad 4.25$$

Without any loss of generality we may assume that the electric field $\boldsymbol{\mathcal{E}}$ is applied in, say, the x direction. Then 4.25 simplifies into

$$f = f_0 - \frac{q\tau\mathcal{E}_x}{\hbar}\frac{\partial E(k_x)}{\partial k_x}\frac{\partial f_0}{\partial E} = f_0 - \frac{q\tau\mathcal{E}_x}{\hbar}\frac{\partial f_0}{\partial k_x} \qquad 4.26$$

Equation 4.26 reminds us immediately of a first order Taylor's theorem and therefore 4.26 can be written as

$$f\left(k_x, k_y, k_z\right) = f_0\left(k_x - \frac{q\tau\mathcal{E}_x}{\hbar}, k_y, k_z\right) \qquad 4.27$$

and hence it seems as if the whole f distribution has been displaced rigidly along the k_x axis by $q\tau\mathcal{E}/\hbar$. This is shown in figures 4.2 and 4.3 (for q = –e). What do these diagrams tell us?

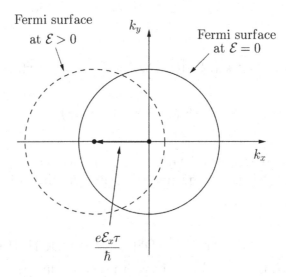

FIGURE 4.2 The action of an electric field is to displace rigidly the equilibrium distribution f_0. This is a side view of the Fermi sphere at T=0.

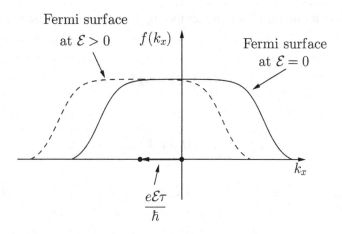

FIGURE 4.3   Same as figure 4.2 but at T>0 and in one dimension.

Exactly what has been assumed in diagrams 3.2a and 3.2b entirely on intuitive grounds, i.e. when a field $\mathcal{E}$ is applied to a system or specimen, the equality of velocities in any direction is perturbed and more electrons **very near the Fermi surface** move in the direction against the field (q<0) than in the opposite direction. In fact, figure 4.3 is a more informative version of figure 3.2b. Furthermore, we have not relied on intuition but we have proved equation 4.27 starting from a rigorous transport equation, the Boltzmann equation. Although we have used the term "displaced rigidly" above, we must emphasize that all the states well below the Fermi surface have not changed their occupancy, everything happens at or near the Fermi surface. In fact, the high conductivity of metals compared to semiconductors is not due to the higher number of electrons in the conduction band but to the much higher electron velocities near the Fermi surface. The velocities well below E_F cancel each other because they always occur in pairs of (+) and (−) directions along the direction of the field $\mathcal{E}$ as the initial diagram 3.2a showed. We can now calculate the current density which will exhibit all the physical characteristics that we have just described.

From 4.25 we get that in 3 dimensions the current density, counting the two spins per electronic state, is

$$J = \frac{2}{V} \sum_{k} (-e) v(k) f(k) \qquad 4.28$$

where in 4.28 the symbol V is the volume of the system and we have explicitly taken care of the negative charge of an electron by writing (−e) for q. The reader may verify that the units of 4.28 are A/m^2. Now we can safely turn the summation into an integration over k but in doing so we will be multiplying the above quantity by $d^3k = dk_x dk_y dk_z$. Since the distance between successive points Δk in the one-dimensional k space is $2\pi/L$, where L is the length of the system, we have to multiply the above quantity by $(L/2\pi)^3$ when turning it from a discrete sum into an integral. Then, remembering that there is no current at equilibrium we get

$$J = -\frac{e}{4\pi^3} \int v(k) f'(k) d^3k \Rightarrow \quad \text{4.29a}$$

$$\Rightarrow J = -\frac{e^2}{4\pi^3} \int v(k)(v(k)\cdot\mathcal{E}) \frac{\tau \partial f_0}{\partial E} d^3k \quad \text{4.29b}$$

Note that in going from equation 4.29a to 4.29b there are 2 minuses involved, one from f' and another from the charge q in f'. The derivative $\partial f_0/\partial E$ is (at $T = 0$) a negative delta function centred at E_F and at higher temperatures it broadens into a Gaussian again centred at E_F. The fact that $\partial f_0 / \partial E$ appears in the expression for the current density J is just another confirmation that conduction takes place near E_F only.

We can now make use of our choice of having one of our axes along $\mathcal{E}$ and define a unit vector in the direction of the electric field by $u = \mathcal{E}/|\mathcal{E}|$. With this definition 4.29 can be written as

$$J = \frac{e^2}{4\pi^3}\left[\int \tau(u\cdot v)^2 \left(\frac{-\partial f_0}{\partial E}\right) d^3k\right]\mathcal{E} \quad \text{4.30}$$

so that the conductivity σ defined by $J = \sigma\mathcal{E}$ (for an isotropic material) is

$$\sigma = \frac{e^2}{4\pi^3} \int \tau(u\cdot v)^2 \left(\frac{-\partial f_0}{\partial E}\right) d^3k \quad \text{4.31}$$

We can simplify 4.30 by substituting $\delta(E(k) - E_F)$ for $-\partial f_0/\partial E$ and $v^2/3$ for $(u\cdot v)^2$. Then we get

$$\sigma = \frac{e^2}{12\pi^3} \int \tau v^2 \delta(E(k) - E_F) d^3k \quad \text{4.32}$$

To further simplify equation 4.32 we need to know how to tackle 3-dimensional integrals in the Brillouin zone as per the one appearing in 4.32. Consider the 3-dimensional k space (the Brillouin zone) depicted in figure 4.4 and consider two surfaces in this space, one corresponding to a constant energy E surface and the other to a $(E + dE)$ surface. The shell of width dE can be divided into infinitesimally small cylindrical elements of base area dS and height dk. The volume of such element is

$$d^3k = dSdk = dS\frac{dE}{|\nabla_k E(k)|} \quad \text{4.33}$$

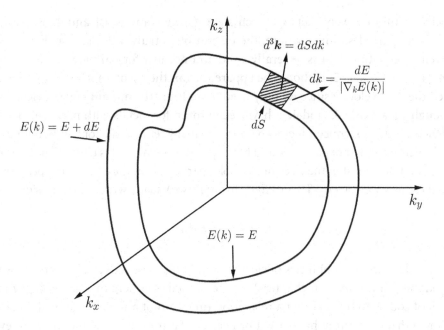

FIGURE 4.4   Construction of a volume element in the 3-dimensional k space.

Using 4.33, the conductivity σ can now be written as

$$\sigma = \frac{e^2}{12\pi^3} \int_{\substack{Fermi \\ surface}} \frac{\tau v^2 \, dS}{\left| \nabla_k E(k) \right|} =$$

$$= \frac{e^2}{12\pi^3 \hbar} \int_{\substack{Fermi \\ surface}} \tau v \, dS \qquad\qquad 4.34$$

A further simplification of 4.34 occurs if we consider electrons in metals in a parabolic band of effective mass m^*. Then using the relations $v_F = \hbar k_F / m^*$ and $k_F^3 = 3\pi^2 n$ of chapter 2 (equation 2.72 may be put into this form) then equation 4.34 reduces to

$$\sigma = \frac{e^2}{12\pi^3 \hbar} \tau_F v_F 4\pi k_F^2 = \frac{ne^2 \tau_F}{m^*} \qquad\qquad 4.35$$

Remembering that $\sigma = ne\mu$ where μ is the mobility we recover the main result of the previous chapter

$$\mu = \frac{e\tau_F}{m^*} \qquad\qquad 4.36$$

But we have done much more than recovering a previous result. We have derived an expression for the conductivity σ and mobility μ, equations 4.34 and 4.36, assuming that the electrons are in wavepackets of Bloch states (see equations 4.1 and 4.2) and are not mere classical particles. Furthermore, the derivation is truly 3 dimensional. The expression given by equation 4.34 is generally valid and not just for parabolic bands of metals when 4.34 reduces to 4.35. It should be apparent from the theory above that we have also obtained the drift velocity since we only need to multiply the mobility by the electric field.

Although equations 4.35 and 4.36 have been proven for metals with parabolic bands, we extend their validity to semiconductors because a) usually semiconductor bands, both conduction and valence bands, are parabolic and b) the pseudo-Fermi level is close to the appropriate band edge and then most of the electrons or holes participate in the conduction process (since most of them lie within a few kT from E_F). Therefore, we usually write the expression

$$\sigma = n|e|\mu_n + p|e|\mu_p$$

where μ_n and μ_p are the mobilities of electrons and holes respectively. Everything we have said about relaxation times in chapter 3 can be carried over here in chapter 4, paying due attention of course to the fact that a relaxation time cannot always be defined. Finally, we have shown that everything happens at or near the Fermi surface, something not evident from chapter 3. In this respect we note that equations 4.35 and 4.36 are slightly misleading as they seem to indicate that all electrons in metals participate in the conduction process.

4.5 DIFFUSION CURRENTS

In the previous chapter we saw that apart from the drift current (which is the result of acceleration by the electric field and collisions) there is also the diffusion current. How does this derive from the Boltzmann equation? It derives from the ∇E_F that we have put to zero. Before proceeding further, the reader is advised to reread the caution suggested by the remarks prior to equation 4.17. Since E_F enters the probability of occupation of a state, it should be obvious that any spatial variation of E_F implies a corresponding variation of the electron density $n(r)$. However, as previously indicated, given that E_F characterizes a system at equilibrium and is unique, it is highly debatable that such concepts rest on firm ground. These issues are discussed in the next chapter and will lead us to the methodology of quantum conduction. However, on the assumption of slow variation of the above quantities within large systems, we can cautiously use such concepts.

Taking into account the term ∇E_F and assuming that in equation 4.23 the magnetic field $\boldsymbol{B} = 0$ and $\nabla T = 0$, we find that we can follow exactly the same steps of the previous section and write the general expression for the current density, equation 4.28, in the form

$$\boldsymbol{J} = eK_0 \left(e\boldsymbol{\mathcal{E}} + \nabla_r E_F \right) \qquad \text{4.37a}$$

where

$$K_0 = \frac{1}{4\pi^3} \int \tau (\boldsymbol{u} \cdot \boldsymbol{v})^2 \left(\frac{-\partial f_0}{\partial E} \right) d^3 k \qquad \text{4.37b}$$

and u is a unit vector in the direction of the electric field $\mathcal{E}$. Taking into account equation 4.31, the above equation can be put into the form

$$J = \sigma\left(\mathcal{E} + \frac{1}{e}\nabla_r E_F\right) \qquad 4.38$$

so we have come to the desired result that in isothermal conditions $(\nabla T = 0)$ we have two types of current densities, a drift current $\sigma\mathcal{E}$ and a diffusion current density $\sigma/e\nabla_r E_F$.

All that remains now is to put the second term in equation 4.38 into the well-known form of Fick's law. Again let us assume a metal with a parabolic conduction band of effective mass m^*. In this case we have from chapter 2, equation 2.72, that

$$E_F = \left(\frac{\hbar^2}{2m^*}\right)\left(3\pi^2 n\right)^{2/3}$$

from which we can deduce that

$$\frac{\nabla_r E_F}{E_F} = \frac{2}{3}\frac{\nabla_r n}{n} \qquad 4.39$$

so that 4.38 can be written as

$$J = ne\mu\mathcal{E} + eD\nabla_r n \qquad 4.40a$$

where

$$D = \frac{2}{3}\frac{E_F}{e}\mu \qquad 4.40b$$

If, on the other hand, we have a semiconductor, from the exponential relation between electron density and Fermi level we get

$$\frac{\nabla_r n}{n} = \frac{\nabla_r E_F}{KT}$$

where K = Boltzmann's constant. Then equation 4.40a remains valid but then

$$D = \frac{KT}{e\mu}$$

Our analysis so far, which was based on the classical Boltzmann equation, has put on a firm basis all of the results of the previous chapter which were based on the mechanics of classical particles without regard to the existence of Bloch states and $\boldsymbol{k}$ space. The Boltzmann equation is also the basis for understanding many solid state phenomena with technological importance such as the Peltier effect. Since our final goal is the quantum analysis of nanoelectronic devices, we leave the description of some of these effects as problems at the end of the chapter. We analyze however two such important effects, the application of a magnetic field or Hall effect and the application of a temperature gradient or the Seebeck effect. Before we do that, we obtain a general expression for the current density in the presence of both an electric field and a thermal gradient.

4.6 GENERAL EXPRESSION FOR THE CURRENT DENSITY

If equation 4.18b instead of 4.18a is substituted in 4.16 we get for Φ, assuming $B = 0$, the following equation (instead of equation 4.21)

$$\Phi(\boldsymbol{k},\boldsymbol{r}) = \tau\left[-e\boldsymbol{\mathcal{E}} - T\nabla_r\left(\frac{E_F}{T}\right) - E\nabla_r ln(T)\right]\cdot\boldsymbol{v} \qquad 4.41$$

Note again that in the above equation we have written q = −e. Using the general expression for the current density, equation 4.28, and following the same steps as in equations 4.28–4.30, we find that for the current density in the presence of a temperature variation we can write

$$\boldsymbol{J} = eK_0\left[e\boldsymbol{\mathcal{E}} + T\nabla_r\left(\frac{E_F}{T}\right)\right] + eK_1\nabla_r lnT \qquad 4.42a$$

where

$$K_1 = \frac{1}{4\pi^3}\int\tau(\boldsymbol{u}\cdot\boldsymbol{v})^2(E)\left(-\frac{\partial f_0}{\partial E}\right)d^3\boldsymbol{k} \qquad 4.42b$$

4.7 APPLICATION OF A THERMAL GRADIENT, THE SEEBECK EFFECT

Equation 4.42a may be written after some standard manipulations involving differentiation of products of functions

$$\boldsymbol{J} = e^2K_0\left[\boldsymbol{\mathcal{E}} + \frac{1}{e}\nabla E_F - S(T)\nabla T\right] \qquad 4.43a$$

where

$$S(T) = \frac{1}{-eT}\left(\frac{K_1}{K_0} - E_F\right) \qquad 4.43b$$

is called the Seebeck coefficient. Now we consider an open circuited bar with a temperature difference at its two ends as in figure 4.5. No electric field is applied, but an electric field

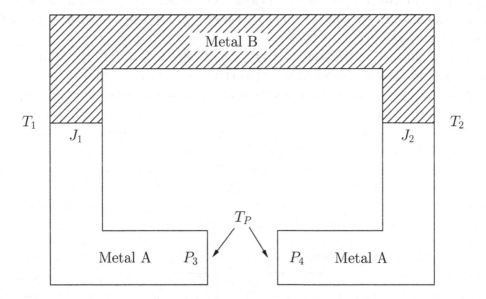

T_1 _____ $\mathcal{E}$ _____ $T_2 > T_1$

FIGURE 4.5   A temperature difference $T_2 - T_1$ creates an electric field in a bar, as shown.

develops in the bar because electrons move in a transient condition from the hot end to the cold end. At steady state, however, J will be zero because we have an open circuit. Then

$$\mathcal{E} = -\frac{1}{e}\nabla E_F + S(T)\nabla T \qquad\qquad 4.44$$

As a result of the movement of the carriers and the establishment of the electric field $\mathcal{E}$ in the bar, a potential difference ΔV appears between the ends of the bar. This is the Seebeck effect. A slightly more complicated experimental configuration than that of figure 4.5 is needed to measure this difference as is shown in figure 4.6. The two junctions $J1$, $J2$ between the two metals A and B are kept at different temperatures T_1 and T_2. At some point anywhere in metal A away from the two junctions a small cut is made creating two end points, call them P3 and P4, between which the potential difference ΔV is measured. The points P3 and P4 are kept at the same temperature T_p. We will have

$$\Delta V = -\int_{P3}^{P4}\mathcal{E}\cdot dl = \frac{1}{e}\int_{P3}^{P4}\nabla E_F dl - \int_{P3}^{P4}S(T)\nabla T dl \qquad\qquad 4.45$$

The line integrals can be taken along any path joining P3 to P4.

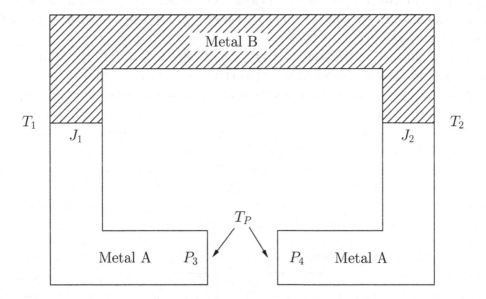

FIGURE 4.6   Experimental set-up for the measurement of the Seebeck coefficient.

Since the temperature at P3 and P4 is the same (T_p), the line integral in the RHS of equation 4.45 involving the gradient of E_F is zero. For the line integral involving the Seebeck coefficient we will have

$$\int_{P3}^{P4} S(T)\nabla T dl = \int_{P3}^{J1} S_A(T)dT + \int_{J1}^{J2} S_B(T)dT + \int_{J2}^{P4} S_A(T)dT =$$

$$= \int_{T_p}^{T_1} S_A(T)dT + \int_{T_1}^{T_2} S_B(T)dT - \int_{T_p}^{T_2} S_A(T)dT \Rightarrow$$

$$\Rightarrow \Delta V = \int_{T_1}^{T_2} S_A(T)dT - \int_{T_1}^{T_2} S_B(T)dT \qquad 4.46$$

Equation 4.46 is actually used to measure the Seebeck or thermoelectric power coefficient of a metal. The material lead is used as metal A which has a negligible Seebeck coefficient and then ΔV is measured so as to obtain $S_B(T)$.

4.8 SATURATION OF DRIFT VELOCITY

Extensive mobility experiments have shown that the velocity does not increase continuously with an increasing electric field $\mathcal{E}$ but actually saturates at a constant value v^{sat} as shown in figure 4.7 (1D model). The low field part of the curve is discussed in section 4.4 where the whole electron distribution $f(k_x)$ shifts rigidly along the k_x axis (along which $\mathcal{E}$ is applied) by an amount proportional to the electric field $\mathcal{E}$. But at high $\mathcal{E}$ it seems as if the electric field is unable to change $f(k)$. We have to look and find out what particular assumptions that we have made in sections 4.3 and 4.4 are no longer valid and no longer give a drift velocity proportional to $\mathcal{E}$.

Before we do that, we present the physical processes that lead to the saturation of velocity. As the electron is accelerated and its energy increases, it interacts strongly with the lattice and gives up the energy acquired by the field to the lattice in the form of optical

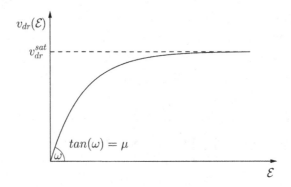

FIGURE 4.7 Saturation of drift velocity at high electric fields.

phonons, a mode of lattice vibrations(see Appendix B). It should then be obvious that the collisions of the electrons with the lattice are no longer elastic and the assumption of the equality of the microscopic transition rates $P(k,k')$ and $P(k',k)$ (see equation 4.9) is no longer valid. In short, from a certain value of the electric field onwards, the extra kinetic energy of the electron is given up to the lattice and the electron velocity stays constant. The description of this phenomenon through the Boltzmann equation is beyond the scope of this book. It must be emphasized that when the collisions are inelastic even the relaxation time approximation is no longer valid. We note, however, that one can define a mean time between inelastic collisions, and we are certainly going to make use of this concept in the next chapter, but one cannot use it to approximate the collisions term in the Boltzmann equation. An empirical relation which reproduces the above discussed effect is

$$v_{dr}(\mathcal{E}) = \frac{\mu\mathcal{E}}{1+\dfrac{\mu\mathcal{E}}{v_{dr}^{sat}}} \tag{4.47}$$

where μ is the low-field mobility.

4.9 GUNN EFFECT AND VELOCITY OVERSHOOT

The drift velocity does not always saturate in the way shown in figure 4.7. In particular, it may not show a monotonic behaviour but it may reach a maximum and then saturate at a value below the maximum. The phenomenon does not occur in all semiconductors but only in the ones that have a direct gap with a higher subsidiary conduction band minimum as in GaAs see figure 2.11. The variation of the drift velocity in GaAs is shown schematically in figure 4.8. The analysis can be performed using a simple particle approach and does not need the Boltzmann equation which has, so far, verified any particle picture that we have used.

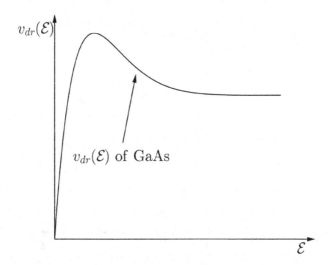

FIGURE 4.8   Variation of drift velocity with electric field in a semiconductor with a direct band-gap.

At small electric fields $\mathcal{E}$ all the electrons in the conduction band reside in the conduction band minimum at $k = 0$. As the electric field $\mathcal{E}$ increases, the average or drift velocity of these electrons increases in accordance with what we have described in section 4.4 (or chapter 3). But when $\mathcal{E}$ reaches the order of magnitude of a few kV/cm, some electrons have gained enough energy to make the jump to the subsidiary conduction band minimum at $k \neq 0$ which, however, has a heavier effective mass than the one at $k = 0$. As $\mathcal{E}$ is further increased, more and more electrons jump to this heavier effective mass minimum and hence the average velocity falls. Any further increase of $\mathcal{E}$ simply leads to saturation. This non-monotonic dependence of velocity on the electric field gives rise to the Gunn effect (discovered by Gunn at IBM), where the current density J as a function the electric field $\mathcal{E}$ follows a variation with $\mathcal{E}$ similar to the $v_{dr}(\mathcal{E})$ of figure 4.8. This is shown in figure 4.9. As can be seen from this figure, the Gunn effect displays a region of negative differential resistance for some $\mathcal{E}$ which can be exploited for engineering microwave oscillators.

A very simple particle-like model can give a mathematical description of the Gunn effect. Let the number of electrons per unit volume n in the conduction band of GaAs, at any time, be divided into those in the light effective mass band-edge (i.e. minimum) n_l and those in the heavy mass band-edge n_h. Then

$$n_l + n_h = n \equiv \text{a constant}$$

Obviously for the derivatives with respect to $\mathcal{E}$ we will have

$$\frac{dn_l}{d\mathcal{E}} = -\frac{dn_h}{d\mathcal{E}} \qquad\qquad 4.48$$

Differentiating the basic equation $J = \sigma\mathcal{E}$ we have

$$\frac{dJ}{d\mathcal{E}} = \sigma + \mathcal{E}\frac{d\sigma}{d\mathcal{E}} \qquad\qquad 4.49$$

since σ is $\mathcal{E}$ dependent because n_l and n_h are functions of $\mathcal{E}$.

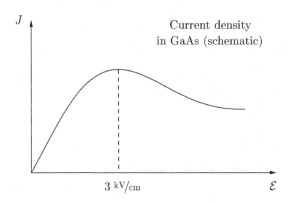

FIGURE 4.9 Schematic variation of current density with electric field in GaAs.

But

$$\frac{d\sigma}{d\mathcal{E}} = e\mu_l \frac{dn_l}{d\mathcal{E}} + e\mu_h \frac{dn_h}{d\mathcal{E}} \Rightarrow$$

$$\Rightarrow \frac{d\sigma}{d\mathcal{E}} = -e\mu_l \frac{dn_h}{d\mathcal{E}} + e\mu_h \frac{dn_h}{d\mathcal{E}} =$$

$$= e\frac{dn_h}{d\mathcal{E}}[\mu_h - \mu_l] \qquad\qquad 4.50$$

where μ_l and μ_h are the mobilities of the light mass minimum and of the heavy mass minimum respectively. Therefore, the RHS of 4.50 is negative $(\mu_h < \mu_l)$ and can make $\dfrac{dJ}{d\mathcal{E}}$ negative in 4.49. This simple model gives the decreasing with $\mathcal{E}$ part of the current density versus electric field characteristic. It should be clear that a more complicated empirical relation than 4.47 is needed to describe the velocity–field relation of this non-monotonic case.

So far what we have presented pertains to steady state conditions, i.e. independent of time. However, an interesting case arises in transient conditions called velocity overshoot. If the path of an electron is short or alternatively scattering mechanisms are absent during a short section of the electrons path, the electron may reach a velocity above its steady state value before it relaxes to its steady state velocity. Such effects are very important for modern state of the art transistors.

4.10 THE (CLASSICAL) HALL EFFECT

So far we have not investigated the effect of a magnetic field on the transport properties of a medium and its effect on the distribution function $f(\mathbf{k}, \mathbf{r})$. The Hall effect, i.e. the effect of the application of a magnetic field on a medium, plays a central role in semiconductor technology since it is used both as a sensor and as a means of detecting the polarity of the carriers. We have left it to the end of this chapter to show that under well behaved conditions the particle description and the Boltzmann equation render the same result. However, this is not always true, there is also the Quantum Hall effect, not describable in particle terms. We will give both pictures, first the particle one (usually taught at undergraduate level) and then the one usingthe Boltzmann equation.

Consider a rectangular long wire of width w and depth d, as in figure 4.10, along the length of which a potential V_y is applied. A magnetic field B_z is also applied perpendicularly to its xy surface pointing upwards. Had there not been B_z, the electrons would have traveled along the positive y direction but because of B_z the carriers are curved along x, as shown in figure 4.10 by the magnetic force $\mathbf{F}_B = -e\mathbf{v} \times \mathbf{B}$. At steady state, the accumulated electrons on one side of the width of the wire will create an electric field $\mathcal{E}$ which will counterbalance the effect of B_z and we will then have zero force on the electrons.

$$\mathbf{F} = -e[\boldsymbol{\mathcal{E}} + \mathbf{v} \times \mathbf{B}] \Rightarrow 0 = \mathcal{E}_x + v_y B_z \qquad\qquad 4.51$$

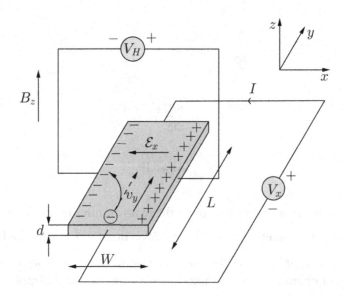

FIGURE 4.10   Set-up for the measurement of the Hall effect; see text for the directions of the fields shown.

$\mathcal{E}$ is negative since $v_y B_z > 0$. Since the charge carriers are electrons (i.e. we have an N – type semiconductor) their velocity is along the $+y$ direction (towards the positive pole) and $v \times B$ points to the $+x$ direction so that $F_B = -ev \times B$ points towards the $-x$ direction and electron accumulation will occur on the left side of the specimen, producing an electro-static voltage difference, the Hall voltage V_H as shown. Had the carriers been holes the polarity of V_H would have been the opposite. Hence the Hall effect may be used as a method of detecting the polarity of the carriers by just looking at V_H. Equation 4.51 together with $\left|\mathcal{E}_x\right| = \left|\dfrac{V_H}{W}\right|$ will give

$$V_H = v_y B_z w \qquad\qquad 4.52$$

Now the current is

$$I_y = wd(en)v_y \qquad\qquad 4.53$$

Substituting for w we get

$$V_H = \frac{B_z I_y}{end} \qquad\qquad 4.54$$

The critical reader must have seen the hidden assumption in deriving 4.54, the assumption that the induced charge is like a surface or sheet-like charge residing on the edges of the rectangular wire. We therefore proceed to give a more proper analysis based on the

Boltzmann equation. In section 4.3 we obtained a final form for this equation, equation 4.23. It is actually easier however if we work from its previous form, equation 4.21.

Assuming no temperature variation across the sample, we get from 4.21 that the deviation from equilibrium is

$$f'(k) = -\varphi \frac{\partial f_0}{\partial E} = -q\tau \mathcal{E} \cdot v(k) \frac{\partial f_0}{\partial E} - \frac{q}{\hbar} \tau (v(k) \times B) \nabla_k f' \qquad 4.55$$

The first term of the RHS of equation 4.55 is exactly what one would get if there were only an electric field $\mathcal{E}$ present. Guided by this, we try a solution of the form

$$f'(k) = -q \frac{\partial f_0}{\partial E} \tau v(k) \cdot A \qquad 4.56$$

where A is to be determined.

Assuming that we can write $m^* v(k) = \hbar k$ we substitute equation 4.56 in both the LHS and RHS of equation 4.55 and we get

$$-q \frac{\partial f_0}{\partial E} \tau v(k) \cdot A = -q\tau \mathcal{E} v(k) \frac{\partial f_0}{\partial E} + \frac{q\tau}{\hbar} (v(k) \times B) \cdot \left(\frac{q \partial f_0}{\partial E} \frac{\tau \hbar A}{m^*} \right) \qquad 4.57$$

Deleting $\left(-q\tau \frac{\partial f_0}{\partial E} \right)$ from both sides and rearranging we get

$$v(k) \cdot \mathcal{E} = v(k) \cdot A + \frac{q\tau}{m^*} (v(k) \times B) \cdot A$$

or

$$\mathcal{E} = A + \frac{q\tau}{m} (B \times A) \qquad 4.58$$

Solving 4.58 for A will give $f'(k)$ and this will give in turn the currents of the Hall effect. However, we can follow a shorter path towards this goal.

We observe that expression 4.56 is of the same form as 4.25. Following the same steps as those of equations 4.25 to 4.31 (which we do not need to repeat here) we get

$$J = \sigma A \qquad 4.59$$

where σ is the conductivity of the sample in the Hall effect assuming an isotropic medium. Substituting back in 4.58 we have

$$\mathcal{E} = \frac{1}{\sigma} J + \frac{q\tau}{m^* \sigma} B \times J \qquad 4.60$$

This equation describes fully the Hall effect. It tells us that at steady state the electric field needed to create the current density $J = J_y$ is not solely along y but has two components, one $\mathcal{E}_y = J_y/\sigma$ and a perpendicular to this one $\mathcal{E}_x = \dfrac{e\tau}{m^*\sigma}B_z J_y$. It is left as a simple exercise to the student to show that $\mathcal{E}_x$, $\mathcal{E}_y$ are consistent with V_H in equation 4.54 if q=−e.

PROBLEMS

4.1 Show that the transition probabilities $P_{k,k'}$ of equation 4.9 are equal at equilibrium.

4.2 Find a closed form for the relaxation time τ (**k**) when the electric field applied to a solid is producing only elastic collisions.

4.3 The energy flux is defined as

$$U = \frac{1}{4\pi^3}\int E(k)v(k)d^3k$$

Using the general expression for the non-equilibrium distribution f (equation 4.41) obtain an expression for U in the form of equation 4.42.

4.4 Prove that the relationship between U and J is of the form

$$U = AJ - k_\theta \nabla T$$

Find expressions for A,k_θ.

4.5 Prove the Wiedmannn–Franz law that the ratio of the electrical σ to the thermal conductivity k_θ (defined above) is constant.

CHAPTER 5

The Quantum Theory of Conduction

5.1 CRITIQUE OF THE BOLTZMANN EQUATION, REGIMES OF CONDUCTION

It was made clear in the previous chapter that the Boltzmann equation is valid only for particles. Electrons can only be thought as such if they have a wavefunction of the form of a wavepacket made out of Bloch functions, well localized in both the r and the k space. Then the maximum of the wavepacket moves according to Newtonian laws and exhibits particle behaviour with the real mass m replaced by the effective mass m^*. How well the wavepacket should be localized? Obviously the extent of the wavepacket must be much smaller than the size of the system L (in one dimension). The former is usually of the wavelength λ. Based on the above, one would think that we only need to distinguish two regimes $L \gg \lambda$ and $L \leqslant \lambda$. When $L \leqslant \lambda$ any notion of a particle behaviour is meaningless and we are definitely within the regime of quantum transport, but when $L \gg \lambda$ the situation is not so simple and we need an extra characteristic length to describe it. This extra characteristic length is the dephasing length or distance L_φ.

This is the distance an electron has to travel before it suffers an inelastic collision. Note that since elastic collisions are more frequent then inelastic, the electron must have undergone through many elastic collisions before it suffers an inelastic one. The truly macroscopic regime is when $L \gg L_\varphi$. In this case an electron, on its way from one electrode to the other one, suffers many inelastic collisions and for every distance L_φ it travels it loses the memory of its phase, the characteristic quantity which makes it a wave. In this case we have particle behaviour. What happens if $L < L_\varphi$ but we still have $L > \lambda$? To better understand the various regimes we are discussing, we note that L_φ is in the range of a few microns and the range of the wavelength of electrons in semiconductors is nanometers or tens of nanometers. In this case of $[\lambda < L < L_\varphi]$, we have what we call mesoscopic regime and the carriers of electricity, the electrons, preserve their wave nature while experiencing only elastic scattering in their path from one electrode to the other. In both the truly quantum regime and the mesoscopic regime, a wave treatment of conductance is necessary.

Present day electronic devices have channels in the nanometre range. At the time of writing this book, device manufacturers have accomplished the 14nm channel length and are working on the 7nm one. These lengths fall definitely below the values of the mean free path in both Si and GaAs. Then we have what we call ballistic transport. The electron traverses the distance between contacts without any scattering. This phenomenon belongs to both the truly quantum and mesoscopic regimes, but it was observed originally in mesoscopic systems. A serious conceptual problem arises: if there are no collisions, where is the energy supplied by the external voltage dissipated? We will tackle this problem later in Part III where we discuss devices.

Apart from the main problem of dealing with electrons exhibiting wave behavior, another problem occurring with the Boltzmann equation, if it were to be applied to nanosystems, is the definition of a Fermi level which is position dependent. This is rarely discussed. When we derived the Boltzmann equation, we took the derivative of the equilibrium distribution $\partial f_0 / \partial E$ assuming a functional dependence $E_F(r)$ in f_0. However the Fermi level is unique and in principle defined only at equilibrium. How then can this assumption be justified? The essence of such an approximation, as explained in the previous chapter, is that every volume element dV, around point r, can be considered on the one hand as mathematically infinitesimal and on the other hand as physically containing many unit cells in equilibrium with its surroundings, so that a local Fermi level can be defined. Obviously such an approach requires an infinite or very large system. The Landauer formalism that we will describe in the next sections treats electrons strictly as waves and eliminates this latter problem by requiring the existence of only the Fermi levels of the electrodes which can always be considered as thermodynamic reservoirs with a fixed Fermi level.

5.2 ELECTRONIC STRUCTURE OF LOW-DIMENSIONAL SYSTEMS

Most modern electronic devices are considered low-dimensional systems, i.e. systems which have one or more of their dimensions in the range of the wavelength of electrons. Note that the wavelength of the electrons in a given semiconductor depends on its effective mass through the simple relation $\lambda = \lambda_0 \sqrt{\dfrac{m}{m^*}}$, where λ_0 is the wavelength of electrons in vacuum. We therefore have to extend our knowledge of eigenstates and density of states (DOS) for such systems before we begin our discussion of quantum transport. There are 3 types of low-dimensional systems and these are shown in figures 5.1a-c. These are a) the 2-dimensional systems in the form of a slab, see figure 5.1a, b) the 1-dimensional systems in the form of a wire, see figure 5.1b, and c) the 0-dimensional systems in the form of a dot, see figure 5.1c. It should be clear that the numbers 2, 1, and 0 in front of the word dimensional refer to the number of extended (almost infinite) dimensions. Such systems do not stand in vacuum but are realized by conducting media surrounded by dielectric parts or semiconductors of higher band gap than the semiconductor where conduction takes place—see figure 5.2 for such an example.

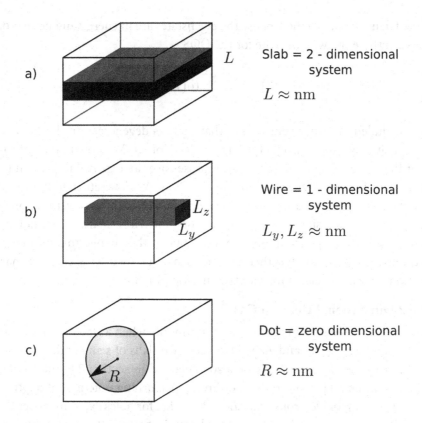

FIGURE 5.1 The 2-, 1-, and 0-dimensional nanosystems.

Prior to calculating the energy spectra of such systems we obtain another useful formula for the density of states (DOS). In chapter 2, the DOS was defined by the relationship 2.51a. We remind the reader that (in the limit $\Delta E \to dE$)

$$g(E)dE = \frac{dN(E)}{V} \qquad\qquad 2.51a$$

where $dN(E)$ is the number of states between E and $E + dE$, V is the volume of the system, and $g(E)$ is the DOS, that is the number of states per unit energy per unit volume. Note

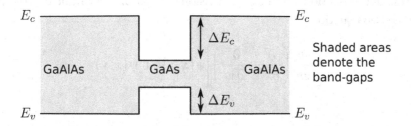

FIGURE 5.2 Creation of a nanolength system by sandwiching a piece of a semiconductor between two semiconductors of a larger band-gap.

that the word density refers to how dense the eigenstates are in energy and not to volume V. For any system therefore we can write for the DOS

$$g(E)=\left(\frac{1}{V}\right)\sum_i(\delta(E-E_i))$$

5.1

The spin is included by either considering that there is degeneracy in the E_i (as we have done) or explicitly by inserting a factor 2 in the RHS of 5.1. We note that as $g(E)$ is integrated over the energy E, everytime the latter passes over an eigenvalue E_i the integration will yield a 1, so 5.1 is a correct expression for the DOS. We now return to the specific low-dimensional systems we were discussing. These systems will exhibit wavefunction localization effects over the dimension in which they are nanometric. Usually we are not interested in the wavefunction change from atom to atom but in the change this localization will effect on the energy spectrum. It is therefore advisable that we use the effective mass equation of section 2.8 which deals only with the envelop function.

5.2a: The 2-Dimensional Electron Gas

As already stated, such a system results most commonly when a slab of a semiconducting material of a band gap E_{g1} is sandwiched between two slabs of semiconductor material of band gap $E_{g2} > E_{g1}$, see figure 5.2. As we also explained in section 2.8, the whole system may be thought as being produced by a confining–perturbing potential of depth equal to $V_{per}^{el} = \Delta E_C = E_{C2} - E_{C1}$ for electrons and $\Delta E_V = E_{V2} - E_{V1}$ for holes extending over the width L of the smaller band gap semiconductor, as shown in figure 5.2. Then assuming planar interfaces, we can write for the envelope function, see figure 5.1a

$$\hat{\phi}(x,y,z)=\psi(x,y)X(z)\text{for }0\leqslant z\leqslant L$$

5.2

The effective mass equation 2.83 for electrons becomes

$$\left[\frac{-\hbar^2\nabla^2}{2m_n^*}+V_{per}^{el}(z)\right]\hat{\phi}(x,y,z)=E\hat{\phi}(x,y,z)$$

5.3

where in 5.3 we measure energies from the bottom of the unperturbed conduction band. Note that, in accordance with our notation for the Schroedinger equation, the symbol V here stands for potential energy, not potential. Using the method of separation of variables that we presented in section 1.3, the 3-dimensional equation 5.3 can be decomposed into the following two equations.

$$\left[\frac{-\hbar^2}{2m_n^*}\left(\frac{\partial^2}{\partial x^2}+\frac{\partial^2}{\partial y^2}\right)\right]\psi(x,y)=\varepsilon_\parallel\psi(x,y)$$

5.4

and

$$\left[\frac{-\hbar^2}{2m_n^*}\frac{\partial^2}{\partial z^2}+V_{per}^{el}(z)\right]X(z)=\varepsilon_\perp X(z)$$

5.5

where

$$E = \varepsilon_{\|} + \varepsilon_{\perp} \qquad\qquad 5.6$$

We have assumed an isotropic mass m^*, but if the material is not isotropic different effective masses $m_{\|}^*$ and $m_{\perp}^*$ should be used in 5.4 and 5.5 respectively.

The solutions of 5.4 are obviously plane waves and we can therefore write for the eigenvalues

$$\varepsilon_{\|}\left(k_x, k_y\right) = \frac{\hbar^2}{2m^*}\left(k_x^2 + k_y^2\right) = \frac{\hbar^2}{2m^*}k_{\|}^2 \qquad\qquad 5.7$$

The solutions of 5.5, on the other hand, are bound states in the confining well of V_{per}^{el} just as those described in chapter 1, section 1.3. The exact mathematical form that these states take depends on the choice of origin of the coordinate system used. However, it suffices to know that they are discreet bound states and hence their eigenvalues can be labeled as $\varepsilon_{\perp}^i$ where (*i*) is the order of the state. The total energy E can be labeled thus as

$$E_i\left(k_x, k_y\right) = \frac{\hbar^2}{2m}\left(k_x^2 + k_y^2\right) + \varepsilon_{\perp}^i \qquad\qquad 5.8$$

A graph of E in $\boldsymbol{k}$ space is shown in figure 5.3.

Since our system is macroscopic in only the x and y directions and nanoscopic in the z direction, a small modification is needed in rewriting the definition 2.51a (reproduced in the beginning of section 5.2). We must actually write for the DOS of 2-dimensional systems

$$g(E)^{2D}\, d(E) = \frac{dN(E)}{S} \qquad\qquad 5.9$$

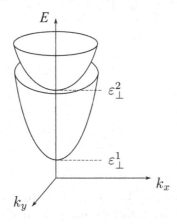

FIGURE 5.3 Energies as a function of the 2-dimensional $\boldsymbol{k}$ vector. The energies $\varepsilon_{\perp}^1$, $\varepsilon_{\perp}^2$ are the discretized energies of the nanosystem.

and

$$g(E)^{2D} = 2/S \sum_{i,k_x,k_y} \delta\left(E - \varepsilon_\perp^i - \frac{\hbar^2}{2m^*}\left(k_x^2 + k_y^2\right) \right) \qquad 5.10$$

where S is the area (in the x, y directions) of the 2-dimensional electron gas and the factor 2 was introduced in equation 5.10 to account explicitly for spin. Let the respective lengths in the x and y directions be L_x and L_y. Then $S = L_x \times L_y$ and the k_x and k_y are, after applying cyclic boundary conditions,

$$k_x = \frac{2\pi n_x}{L_x} \text{ and } k_y = \frac{2\pi n_y}{L_y}$$

where $n_x, n_y = 0, 1, 2, \ldots$

Turning the finite k summations in 5.10 to an integration and remembering to include the factor necessary to do so $\left(\text{which is } S/(2\pi)^2 \text{ here}\right)$ we get

$$g^{2D}(E) = \frac{2}{4\pi^2} \sum_i \iint \delta\left(E - \varepsilon_\perp^i - \frac{\hbar^2}{2m^*}\left(k_x^2 + k_y^2\right) \right) dk_x \, dk_y \qquad 5.11$$

But $dk_x dk_y = kdkd\theta = d\left(\frac{k^2}{2}\right)d\theta$ where θ is the polar angle. Integrating over this angle will yield a factor of 2π, so that we only need to perform the integration over $\pi d\left(k^2\right)$. Equation 5.11 is then transformed into

$$g^{2D}(E) = \frac{1}{2\pi^2} \sum_i \int_0^\infty \delta\left(E - \varepsilon_\perp^i - \frac{\hbar^2 k^2}{2m^*} \right) \pi dk^2 =$$

$$= \frac{m^*}{\pi\hbar^2} \sum_i \int_0^\infty \delta\left(E - \varepsilon_\perp^i - \varepsilon_\parallel \right) d\varepsilon_\parallel =$$

$$= \frac{m^*}{\pi\hbar^2} \sum_i \Theta\left(E - \varepsilon_\perp^i \right) \qquad 5.12$$

where $\Theta(x)$ is the step function (or Heaviside function) defined as follows: $\Theta(x) = 1$ for $x > 0$ and $\Theta(x) = 0$ for $x < 0$. If the 2-dimensional semiconductor is made of Si, the result above has to be multiplied by the generacy of the valleys in Si, f_v, if the energies $\varepsilon_\perp^i$ are treated as non-degenarate.

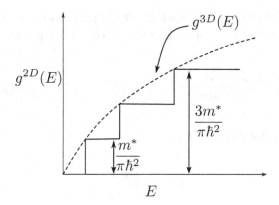

FIGURE 5.4   Density of states of a 2-dimensional nanosystem.

A graph of $g^{2D}(E)$ is shown in figure 5.4, which exhibits a staircase shape because as the energy E increases and moves over an eigenvalue $\varepsilon_\perp^i$ of the well, it gives a contribution $\left(m^*/\pi\hbar^2\right)$ everytime it does so. Note that as the width of the well L in the z direction increases the energy distance between the various $\varepsilon_\perp^i$ decreases and the form of the curve in figure 5.4 approaches a parabola as we expect from a 3D system. 2-dimensional systems are technologically very important because many devices are of this form whereas the very latest generation of transistors are of the one-dimensional systems. We proceed to examine these systems now.

5.2b: The 1-Dimensional Systems or Quantum Wires

1-dimensional systems are equally important to 2-dimensional systems. Most state of the art transistors are 1-dimensional systems. The easiest way to produce such systems is to shrink the length of, say, the x direction which forms one dimension of a 2-dimensional system. This is actually the path that the microelectronics industry has followed for many years in pursuit of Moore's law. A note of warning: conduction may not necessarily take place along an extended direction in many transistor configurations. In all cases discussed above a confining potential is "applied" which derives from the band–edge offset of a larger band gap semiconductor or a dielectric.

We consider the x direction to be the extended one and confinement to occur in the y and z directions, see figure 5.1b. Then we can write for the perturbing potential producing the 1-dimensional system (dropping the superscript el)

$$V_{per}(y,z)=\begin{cases} 0 & \text{for } 0\leqslant y\leqslant L_y \text{ and } 0\leqslant z\leqslant L_z \\ V_b & \text{otherwise} \end{cases} \qquad 5.13$$

We can use the effective mass equation and write for the envelop wavefunction again

$$\hat{\phi}(x,y,z)=\psi(x)X(y,z) \qquad 5.14$$

where $\psi(x)$ are one–dimensional plane waves and $X(y,z)$ denotes localized wavefunctions bounded by the potential of 5.13 above. The $X(y,z)$ obey the equation

$$\left[-\frac{\hbar^2}{2m^*}\left(\frac{\partial^2}{\partial y^2}+\frac{\partial^2}{\partial z^2} \right)+V_{per}(y,z) \right]X_{l,m}(y,z)=\varepsilon_{l,m}X(y,z) \qquad 5.15$$

where $\varepsilon_{l,m}$ relates to the total energy E by

$$E=\varepsilon_{l,m}+\frac{\hbar^2 k_x^2}{2m^*} \qquad 5.16$$

Note that since the system is bound in 2 dimensions the energies $\varepsilon_{l,m}$ and the wavefunctions $X_{l,m}(y,z)$ of the localized states, must be labeled by 2 indices. Note also that again an isotropic effective mass m^* has been used in 5.15. The density of states of a 1-dimensional system $g^{1D}(E)$ can now be written in analogy with 5.9 in the form

$$g^{1D}(E)dE=\frac{dN^{1D}(E)}{L_x} \qquad 5.17$$

where L_x is the length of the system in the extended x direction.

Adapting the general formula 5.1 to 1 dimension we get

$$g^{1D}(E)=\left(\frac{2}{L_x} \right)\sum_{l,m}\sum_{k_x=0}^{\infty}\delta\left(E-\varepsilon_{l,m}-\frac{\hbar^2 k_x^2}{2m^*} \right)= \qquad 5.18a$$

$$=\sum_{l,m}\left(\frac{2}{2\pi} \right)\int_{-\infty}^{\infty}\delta\left(E-\varepsilon_{l,m}-\frac{\hbar^2 k_x^2}{2m^*} \right)dk_x= \qquad 5.18b$$

$$=\frac{2}{\pi}\sum_{l,m}\int_{0}^{\infty}\delta\left(E-\varepsilon_{l,m}-\frac{\hbar^2 k_x^2}{2m^*} \right)dk_x \qquad 5.18c$$

Note that in the equations above, the extra factor 2 in going from 5.18b to 5.18c is due to the change of the limits of integration. Changing the variable of integration $y=\frac{\hbar^2 k_x^2}{2m^*}$ we obtain after some trivial operations

$$g^{1D}(E)=\frac{1}{\pi}\sum_{l,m}\frac{\sqrt{2m^*}}{\hbar}\frac{1}{\sqrt{E-\varepsilon_{l,m}}}\Theta(E-\varepsilon_{l,m}) \qquad 5.19$$

The graph of $g^{1D}(E)$ is shown in figure 5.5.

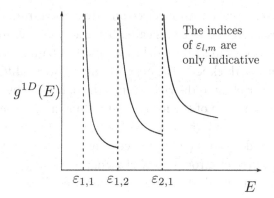

FIGURE 5.5 Density of states of a 1-dimensional system, $\varepsilon_{i,j}$ are the discretized energies of the nanosystem.

5.3 THE LANDAUER FORMALISM

Landauer began tackling the problem discussed in section 5.1 as early as 1957. By now his method is considered the most widely used, if not the most powerful. Concerns have been raised as to the application of this method to conduction of electrons suffering inelastic collisions, but we will not deal with such issues since the continuous shrinking of devices has made such events rare. If one could summarize Landauer's work in one sentence, one would say that if electrons are treated as waves then the conductive properties of a system are essentially its wave transmission properties. We will first tackle a strictly 1-dimensional problem to show this, and then we will generalize to 3-dimensional systems where conduction occurs still along 1 dimension but the device is a 3-dimensional object with 1 or 2 of its dimensions in the nanometric range.

Landauer envisioned a channel connecting two thermodynamic reservoirs which are kept at a constant potential (not potential energy) difference V, see figure 5.6. Then for the Fermi level difference of the reservoirs (which represent the leads to the device) we have $eV = \Delta E_F$. For the leads or electrodes to behave as thermodynamic reservoirs, they must be large enough

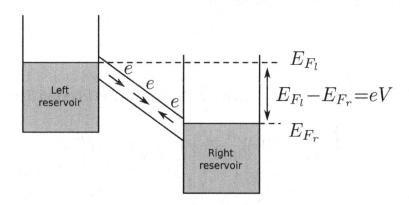

FIGURE 5.6 Conduction as envisaged by Landauer: electrons flow along a channel between two thermodynamic reservoirs separated by an energy eV, where V is the applied voltage.

to keep their Fermi levels constant while electrons are ejected from the left reservoir and thrown to the right reservoir and vice versa after being accelerated through the channel.

We first consider a very simplified model in which the channel can be thought of as a wire of negligible radius or thickness. The net current will be the difference of the current from the left reservoir traveling to the right minus the opposite one, see figure 5.6. Every electron is considered as a wave of the form e^{ikx} traveling either from left to right with a positive wavenumber k or from right to left with a negative k. The current carried by one electron can be calculated by multiplying the current density $J = nev$ by an infinitesimal cross section A, where n=the electron density of a single electron is f/AL with f the Fermi–Dirac distribution. Then this product must be summed over all electrons, i.e. over all occupied states. We have

$$I = I_{lr} - I_{rl} \Rightarrow$$

Assuming that all the electrons emerging from the reservoir reach the other end

$$I = (2e/L)\left[\sum_{k>0} v(k) f_l(k) - \sum_{k>0} v(k) f_r(k)\right]$$ 5.20a

where $v(k)$ is the velocity of the electrons in the channel and f_l, f_r are the equilibrium Ferni–Dirac distributions in the left and right reservoir respectively and the factor 2 comes from the spin of the electron. Note that the second term should actually have $k < 0$ in the summation, but we have taken care of this by the minus in front of the summation. Turning the sums into integrals (i.e. multiplying by (L/2π)) we get

$$I = \frac{e}{\pi}\left[\int_0^\infty v(k) f_l(k)dk - \int_0^\infty v(k) f_r(k)dk\right]$$ 5.20b

As usual we can assume parabolic bands $E(k)$ so that

$$v(k) = \frac{1}{\hbar}\frac{dE(k)}{dk} = \frac{\hbar k}{m^*}$$

Using the above relation and changing the variable of integration from k to E we get

$$I = e\int_0^{E_{Fl}} \frac{1}{\hbar\pi}\frac{dE}{dk} f_l(E(k))\frac{dk}{dE}dE - e\int_0^{E_{Fr}} \frac{1}{\hbar\pi}\frac{dE}{dk} f_r(E(k))\frac{dk}{dE}dE$$

$$= e\int_{E_{Fr}}^{E_{Fl}} \frac{1}{\hbar\pi}[f_l(E) - f_r(E)]dE$$ 5.21

Hence assuming conduction at the Fermi level and $T \rightarrow 0$ the current is (changing $\hbar$ to h)

$$I = \frac{2e}{h}\left(E_{Fl} - E_{Fr}\right) \qquad\qquad 5.22$$

which becomes

$$I = \frac{2e^2}{h}V \qquad\qquad 5.23$$

Note the change from $\hbar$ to h in going from 5.21 to 5.22 to transform to the original result. Equation 5.23 means that the conductance of the wire G defined by $I = GV$ is

$$G = \frac{2e^2}{h} = 12.9 mS \qquad\qquad 5.24$$

Had we not assumed that all electrons emerging from the reservoirs reach the opposite reservoir a factor $T(E)$ equal to the fraction of electrons transmitted through the channel, would have appeared in the equations leading to 5.24 and we would then get

$$G = \frac{2e^2}{h}T(E_F) \qquad\qquad 5.25$$

on the assumption that again conduction occurs very near E_F (as we have seen in the previous chapter) and that $T(E)$ can be represented by an average, equal to $T(E_F)$. This assumption will be lifted immediately below and a more rigorous proof will be given. A further assumption is that the transmission coefficient from left to right $T^{\rightarrow}(E)$ and the transmission coefficient from right to left $T^{\leftarrow}(E)$ are equal. We will not prove this. Note that we did not have to assume anything about the Fermi level in the channel, which in any case is not properly defined. Only the Fermi levels of the reservoirs enter the derivation which are properly defined. A final point, that we wish to make, is that due to the low dimensional character of the system the conductivity is not properly defined and we can only talk about the conductance of the system.

We will now deal with the realistic case, where the conductor – channel is a three dimensional object that has width and depth see figure 5.7a. To simplify matters, we will use Cartesian geometry and coordinates. At thermodynamic equilibrium there is a common Fermi level for both reservoirs and the channel. When a positive voltage difference V is applied between the two leads – reservoirs the Fermi level of the right reservoir is lowered by eV and a net flow of electrons from the left one to the right one is established. Let this voltage difference V produce a potential energy distribution $U(x, y, z)$ which can be separated in parallel (along the channel) and perpendicular components, see figure 5.7b. Note that since we will need both the terms potential

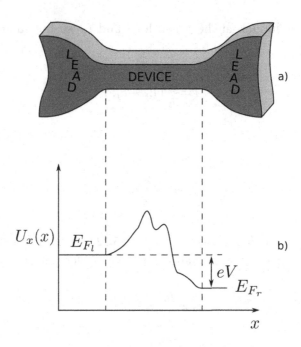

FIGURE 5.7 Same as 5.6 but in three dimensions; (a) the reservoirs are the leads and the channel is the device, and (b) the variation of potential energy along the channel and in the reservoirs-leads.

energy and voltage or potential, we assign now the symbol U to the former and V to the latter.

$$U(x,y,z)=U_x(x)+U_{yz}(y,z) \qquad 5.26$$

Then the wavefunction can be put into a product form in accordance with the discussion in section 5.2

$$\Psi(x,y,z)=\Psi_x(x)\Psi_{yz}(y,z) \qquad 5.27$$

The wavefunctions appearing in equation 5.27 are, of course, the envelop functions. As we have described in section 5.2b, the eigenvalues of this low-dimensional system can be written as $E(k_x,l,m)$, see equation 5.16.

Let f_L and f_R be the number of electrons of a given spin in the left and right reservoir respectively in a given state $E(k_x,l,m)$. Then f_L and f_R are given by the usual Fermi–Dirac function. We assume that the Fermi level of the left reservoir E_F is unaltered and the Fermi level of the right reservoir is lowered by eV. Furthermore the corresponding densities are $2f_L/Vol$ and $2f_R/Vol$ where Vol is the volume of our channel.

Again as in the simplified 1-dimensional case, the current (not current density) is the difference of currents ejected from the left and right reservoirs or algebraically

$$I=I_{LR}+I_{RL}, \text{ with } I_{RL}<0 \qquad 5.28$$

Let E_F denote the common unperturbed Fermi level then the Fermi level of the left reservoir is E_F and of the right reservoir $E_F -$ eV. We have for I_{LR}, following the steps of the 1-dimensional case,

$$I_{LR} = \frac{2e}{L} \sum_{l,m} \sum_{k_x>0} v(k_x) T(E_x) f_L \left(E(k_x,l,m) - E_F \right) \qquad 5.29$$

and for I_{RL} similarly

$$I_{RL} = \frac{2e}{L} \sum_{l,m} \sum_{k_x<0} v(k_x) T(E_x) f_R \left(E(k_x,l,m) - E_F + eV \right) \qquad 5.30$$

In both 5.29 and 5.30 we have included a factor of 2 for spin. As in the elementary case treated above we have divided by L only (instead of Vol) because the cross section A has been absorbed in the current I. Note also that the transmission coefficient depends only on the energy of a particle normal to the barrier which in this case is E_x. Furthermore $k_x > 0$ in 5.29 picks up only the electrons travelling from left to right and $k_x < 0$ in 5.30 picks up only the electrons travelling from right to left.

The total current is according to 5.28

$$I = \frac{2e}{L} \sum_{l,m} \sum_{k_x>0} v(k_x) T(E_x) \left[f_L \left(E(k_x,l,m) - E_F \right) - f_R \left(E(k_x,l,m) - E_F + eV \right) \right] \qquad 5.31$$

Note that in 5.31 we have reversed the sign of the inequality $k_x < 0$ so as to have $v(k_x)$ as a common factor. This amounts to having a minus sign in 5.28 with both currents being positive.

We can now change the order of the summations in 5.31 and perform the sum over l, m first, taking $v(k_x)$ and $T(E_x)$ as common factors since they do not depend on (l,m). Then we will get factors in 5.31 of the form

$$\sum_{l,m} \frac{1}{1 + exp\left(E(k_x,l,m) - E_F \right)/ KT} \equiv N_L(E_x - E_F)$$

and of the form

$$\sum_{l,m} \frac{1}{1 + exp\left(E(k_x,l,m) + eV - E_F / KT \right)} \equiv N_R(E_x - E_F + eV)$$

Hence equation 5.31 for the current becomes

$$I = \frac{2e}{L} \sum_{k_x} v(k_x) T(E_x) \left[N_L(E_x - E_F) - N_R(E_x - E_F + eV) \right] \qquad 5.32$$

As a final step we can turn the summation over k_x into an integration. Remembering to divide by $\dfrac{2\pi}{L}$ we get

$$I = \frac{e}{\pi}\int_k v(k_x) T(E_x)\big[N_L(E_x - E_F) - N_R(E_x + eV - E_F)\big]dk_x =$$

$$= \frac{e}{\pi\hbar}\int T(E_x)\big[N_L(E_x - E_F) - N_R(E_x + eV - E_F)\big]dE_x \qquad 5.33$$

To reach equation 5.33 we have changed the integration from k_x to E_x. Note that we get the 1-dimensional result if we substitute $\hbar = h/(2\pi)$.

This is the celebrated Landauer's formula in a more general form. A close inspection of the required steps for its derivation show that it is not necessary that the eigenstates denoted by the indices (l,m) are localized in the 2-dimensions perpendicular to the direction of propagation as we initially assumed. They can also be extended states in the y and z directions. Of course the value of N_R and N_L will depend critically on such an issue. In the next part, devices, we will see that the parts of the transistors which correspond to the reservoirs of our abstract so far theory, the source and the drain, always have 1 dimension (the width) large enough to be considered macroscopic.

Taking the limit $T \to 0$, i.e. working at very low temperatures, we get immediately Ohm's law. To see this observe that the functions N_R, N_L become step functions for metals with the discontinuity occurring at the corresponding Fermi level of each reservoir, $\mu_1 = E_F$ for the left one and $\mu_2 = E_F - eV$ for the right one. Denoting by ΔN_{RL} the difference appearing in 5.33 and using a first order Taylor expansion for this difference as a function of V we have

$$\Delta N_{RL} = \left(\frac{\partial N}{\partial\mu}\right)(eV) = -eV\frac{\partial N}{\partial E} = eV\delta(E_F - E) \qquad 5.34$$

The minus in the second step of equation 5.34 arises because the N_L and N_R are functions of $E_x - \mu$. Remembering that the derivative of a step function is a delta function and using 5.34 in 5.33 we get

$$I = \frac{2e^2}{h} T(E_F)V \qquad 5.35$$

A more general expression than 5.35 can be obtained if we assume that $T(E_x)$ is not independent of (l,m), i.e. of the transverse quantum states from which the electrons originate. Then we should write $T_{l,m}(E_x)$ instead of simply $T(E_x)$. Note that the transmission coefficient is still a function of the energy normal to the barrier which is E_x. Then starting from equation 5.31, instead of 5.33, and using equation 5.34 we get

$$I = \frac{2e^2}{h}\sum_{l,m} T_{l,m}(E_F)V \qquad 5.36$$

Hence the conductance – which is defined by $I = GV$ – is

$$G = \frac{2e^2}{h} \sum_{l,m} T_{l,m}(E_F)$$

5.37

5.4 THE EFFECTIVE MASS EQUATION FOR HETEROSTRUCTURES

In analyzing low-dimensional systems by the effective mass equation, we were careful to apply this equation to the calculation of the wavefunctions of the same material, be it a planar layer in a sandwich structure or a wire surrounded by a dielectric. We did not examine the variation of a wavefunction across two materials with a band edge offset ΔE_C or ΔE_V between them. To do so we will need boundary conditions for the wavefunction and its derivative, as we did in chapter 1 for the calculation of the tunneling current of an electron beam hitting a barrier in vacuum. However the boundary conditions for the interface of two materials, see figure 5.8, with different effective masses m_1^* and m_2^* are not the usual ones when it comes to the envelop functions.

Let $X(x)$ denote the envelop function along the direction perpendicular to an interface of two materials at $x = 0$. $X(x)$ is the x-component of the total wavefunction as in equation 5.2. Equality of the envelop wavefunction is still required

$$X(x = 0^-) = X(x = 0^+)$$

5.38

But the relation between the derivatives must be modified. Instead of the usual equality we require

$$\frac{1}{m_1} \frac{dX(x)}{dx}\bigg|_{x=0^-} = \frac{1}{m_2} \frac{dX(x)}{dx}\bigg|_{x=0^+}$$

5.39

Why is this? Basically to conserve current. In chapter 1 we saw that the derivative in a certain direction corresponds to the momentum operator in that direction, so the above relation 5.39 constitutes an equality of velocities and hence current is conserved. In fact,

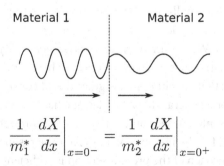

FIGURE 5.8 Matching of a 1-dimensional wavefunction X at an interface of two materials with different effective masses.

the modified boundary condition 5.39 guarantees that the actual wavefunction has a continuous slope when the interface is crossed.

Another complication which arises with heterostructures is the following: in deriving the effective mass approximation (EMA) we have omitted the k dependence of the cellular part u_{nk} of the wavefunction, cf. equation 2.69, keeping only u_{n0}, the cellular part at the band extremum, assumed to be at $k = 0$ as in GaAs. However, when an electron crosses an interface, it may encounter a semiconductor where the band extremum is not at the same k point. Such is the case, for example, for the interface $GaAs/GaP$ and the interface $GaAs/GaAs_{1-x}P_x$ for $x > 0.4$. In this case, the effective mass approximation does not hold, so that great care must be exercised before it is applied.

On the assumption that we have an interface with the band extrema at the same k point we cannot continue to use the EMA of chapter 2 before a further modification in the equation itself is made. We have to write the EMA in the following form (in 1 dimension)

$$\left[\frac{-\hbar^2}{2} \frac{d}{dx} \left[\frac{1}{m^*(x)} \frac{d}{dx} \right] + V(x) \right] X(x) = (E - E_C) X(x) \qquad 5.40$$

In equation 5.40 the meaning of $m^*(x)$ is that it is piecewise constant—it changes from a constant m_1^* to another constant m_2^*. The reason for this mathematical trick is that the hamiltonian of this form preserves its hermiticity. It does not do so if a space varying mass is used in its usual form.

5.5 TRANSMISSION MATRICES, AIRY FUNCTIONS

The prerequisite for the application of the formalism of Landauer is the knowledge of the transmission coefficient of the channel in question. A systematic method of obtaining the transmission coefficient is available if only tunnelling with no scattering is present in the channel. The method consists in a) writing down the form of the wavefunctions in the reservoirs and the channel (which may consist of more than one material), b) matching the wavefunctions and their derivatives at the interfaces, and c) deducing the transmission coefficient as the squared modulus of the ratio of the relevant coefficients relating the incident to the transmitted wave times the ratio of the relevant velocities or wavevectors. We have already encountered this method, though not so formally, in the problem of tunnelling through a rectangular barrier in section 1.7. There, we obtained the transmission coefficient $|G/A|^2$, solving a system of 4 equations in 4 unknowns. We assumed without loss of generality that $A = 1$. Essentially the same method, but using matrix algebra, will be followed here. The method is only useful for a barrier or series of barriers that are either rectangular (i.e. of constant potential) or trapezoidal (i.e. of linear potential). For a general shape barrier one has to revert to the WKB approximation which we initially presented in section 1.8 and a more thorough discussion we will give in the next chapter. However, for the purpose of analyzing the resonant tunneling diode (RTD) in the next section, the approach of the present section is sufficient.

To prepare the ground for the RTD and go one step further than the simple case of the rectangular barrier, we consider the problem of an undoped semiconductor material,

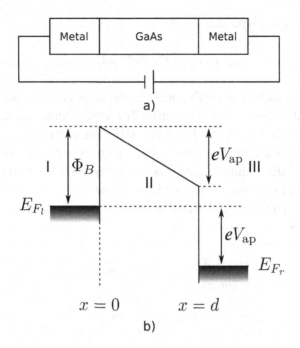

FIGURE 5.9 (a) An intrinsic GaAs sample between two metal leads under an applied voltage V_{ap}, (b) the band structure of the system in the configuration shown in (a).

say GaAs, with two metal or heavily doped semiconductor contacts at its two ends and a voltage difference V_{ap} applied between the contacts, see figure 5.9a. The length of GaAs is d—the energy band diagram corresponding to this arrangement is shown in figure 5.9b. We assume that the whole of the applied voltage is dropped in the semiconductor region. This is justified considering that the latter is undoped and hence has a very high resistance compared to the metal contacts. Furthermore, if we can neglect the small intrinsic concentration n_i in the GaAs semiconductor, then there are no free carriers and the potential variation in the semiconductor is linear according to the Poisson equation. On the other hand, there is no variation of potential energy in the metal contacts if the whole of V_{ap} is dropped in the semiconductor layer.

If the $(+)$ lead of the supply is connected to the right electrode, the Fermi level of layer I is lifted compared to that of layer III or that of layer III is lowered compared to I, see figure 5.9b. We take the origin of energy at the Fermi level of layer III. Then the potential energy in layer I is $U(x) = U_{ap} = (-e)V_{ap}$, a constant. Note that we use the symbol U for the potential energy so that we can keep on using V for the potential or voltage. The eigenstates of layer I can be written as usual as a linear combination of plane waves

$$\Psi_I = A exp\left[i\frac{\sqrt{2m_I^*\left(E-U_{ap}\right)}x}{\hbar}\right] + B exp\left[-i\frac{\sqrt{2m_I^*\left(E-U_{ap}\right)}x}{\hbar}\right] =$$

$$= A exp(ik_I x) + A exp(-ik_I x) \qquad\qquad 5.41$$

The potential energy in layer II is

$$U(x) = U_{ap} + \Phi_B - e\mathcal{E}_x = U_0 - e\mathcal{E}_x \qquad 5.42$$

where Φ_B is the Schottky barrier (in eV) between the metal contact and the semiconductor (cf section 3.10) and $\mathcal{E}_x$ and is the electric field. The Schroedinger equation with a linearly varying potential can be transformed into the Airy differential equation, the solutions of which are the well-known Airy functions $Ai(x)$ and $Bi(x)$ of the first and second kind respectively. The transformation is mathematically elementary but care must be exercised with units. We give this transformation in detail below.

The Schroedinger effective mass equation reads in one dimension in layer II with (for the moment) $m^* = m$

$$\left[\frac{-\hbar^2}{2m} \frac{\partial^2}{\partial x^2} + U_0 - e\mathcal{E}_x x \right] \Psi = E\Psi \Rightarrow$$

$$\Rightarrow \left[+\frac{\partial^2}{\partial x^2} + \frac{2me\mathcal{E}_x x}{\hbar^2} \right] \Psi = \frac{2m(U_0 - E)}{\hbar^2} \Psi \qquad 5.43$$

The first term in the brackets in 5.43 has dimensions of $(length)^{-2}$ and hence so must have the term next to it. Therefore, the factor

$$\frac{\hbar^2}{2me\mathcal{E}_x} \equiv x_0^3$$

must have dimensions $(length)^3$. Hence equation 5.43 transforms into

$$\left[\frac{\partial^2}{\partial \left(\dfrac{x}{x_0} \right)^2} + \frac{x}{x_0} \right] \Psi = \frac{x_0^2 (2m)(U_0 - E)}{\hbar^2} \Psi \qquad 5.44$$

In 5.44 the left-hand side of the equation is dimensionless, so the right-hand side must also be. Hence the factor

$$\frac{x_0^2 \, 2m}{\hbar^2} \equiv (E_0)^{-1}$$

must have the dimensions of $(energy)^{-1}$. We have

$$E_0^{-1} = \frac{2m}{\hbar^2} \left(\frac{\hbar^2}{2me\mathcal{E}_x} \right)^{2/3} \Rightarrow E_0 = \left(\frac{2m}{\hbar^2 e^2 \mathcal{E}_x^2} \right)^{-1/3}$$

Denoting by

$$x' = \frac{x}{x_0}$$

and by

$$E' = \frac{U_0 - E}{E_0}$$

we get

$$\left[\frac{\partial^2}{\partial x'^2} + x'\right]\Psi = E'\Psi$$

Since all variables in front of Ψ are dimensionless, we can define $q = E' - x'$. We then get (using simple instead of partial derivatives)

$$\frac{d^2\Psi}{dq^2} = q\Psi \qquad 5.45$$

This is Airy's 2nd order differential equation, and its particular solutions are the two Airy functions $Ai(q)$ and $Bi(q)$, shown in figures 5.10a and 5.10b respectively. The form that Ψ takes in an infinite triangular well is $Ai(q)$ and is shown in figure 5.10c. The general solution in a finite triangular well is

$$\Psi(q) = C \cdot Ai(q) + D \cdot Bi(q) \qquad 5.46$$

Going back from the mathematical variable q to the physical variables x, Φ_B, U_{ap} and reinstating effective masses we have (noting that $U_{ap} = e\mathcal{E}_x d$)

$$\Psi_{II}(x) = C \cdot Ai\left[\left(\frac{2m_{II}^*}{\hbar^2}\frac{U_{ap}}{d}\right)^{\frac{1}{3}}\left(\frac{d(U_0 - E)}{U_{ap}} - x\right)\right] +$$

$$D \cdot Bi\left[\left(\frac{2m_{II}^*}{\hbar^2}\frac{U_{ap}}{d}\right)^{1/3}\left(\frac{d(U_0 - E)}{U_{ap}} - x\right)\right] =$$

$$= C \cdot Ai\left(\rho_{II}\left(\xi_{II} - x\right)\right) + D \cdot Bi\left(\rho_{II}\left(\xi_{II} - x\right)\right) \qquad 5.47$$

The definitions of ρ_{II} and ξ_{II} are obvious. Finally, in layer III the wavefunction can be written as

$$\Psi_{III} = G exp(ik_{III}x) \qquad 5.48$$

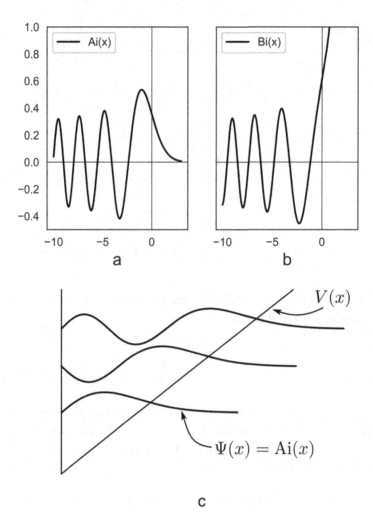

FIGURE 5.10   (a) The Airy function $Ai(x)$ of the first kind and (b) the Airy function of the second kind $Bi(x)$, and (c) the solution of the Schroedinger equation for an infinite length linear quantum well. The first 3 eigenfunctions are shown.

where

$$k_{III} = \frac{\sqrt{2m_{III}^* E}}{\hbar} \qquad 5.49$$

Requiring that the wavefunction $\Psi(x)$ and its derivative $1/md\Psi/dx$ are continuous at the interfaces (according to equations 5.38 and 5.39) we get the following two equations

$$A + B = C * Ai(\rho_{II}\xi_{II}) + D * Bi(\rho_{II}\xi_{II}) \qquad 5.50$$

$$ik_I A - ik_I B = -\frac{m_I^*}{m_{II}^*}\left(C\rho_{II} Ai'(\rho_{II}\xi_{II}) + D\rho_{II} Bi'(\rho_{II}\xi_{II})\right) \qquad 5.51$$

for the boundary at $x = 0$ where Ai' and Bi' are the derivatives of Ai and Bi with respect to x, and the following two

$$C * Ai\left(\rho_{II}\left(\xi_{II} - d\right)\right) + D * Bi\left(\rho_{II}\left(\xi_{II} - d\right)\right) = G \exp\left(ik_{III}d\right) \qquad 5.52$$

$$\frac{m_{III}^*}{m_{II}^*}\left[-C\rho_{II}Ai'\left(\rho_{II}\left(\xi_{II} - d\right)\right) - D\rho_{II}Bi'\left(\rho_{II}\left(\xi_{II} - d\right)\right)\right] = Gik_{III}\exp\left(ik_{III}d\right) \qquad 5.53$$

for the boundary at $x = d$.

The above four equations can be written as two matrix equations of 2×2 size as follows

$$\begin{bmatrix} A \\ B \end{bmatrix} = \frac{1}{2k_I}\begin{bmatrix} k_I & -i \\ k_I & i \end{bmatrix}\begin{bmatrix} Ai\left(\rho_{II}\xi_{II}\right) & Bi\left(\rho_{II}\xi_{II}\right) \\ -\dfrac{m_I^*}{m_{II}^*}\rho_{II}Ai'\left(\rho_{II}\xi_{II}\right) & -\dfrac{m_I^*}{m_{II}^*}\rho_{II}Bi'\left(\rho_{II}\xi_{II}\right) \end{bmatrix}\begin{bmatrix} C \\ D \end{bmatrix} \qquad 5.54$$

and

$$\begin{bmatrix} C \\ D \end{bmatrix} = \begin{bmatrix} Ai\left(\rho_{II}(\xi_{II} - d)\right) & Bi\left(\rho_{II}(\xi_{II} - d)\right) \\ -\dfrac{m_I^*}{m_{II}^*}\rho_{II}Ai'\left(\rho_{II}(\xi_{II} - d)\right) & -\dfrac{m_I^*}{m_{II}^*}\rho_{II}Bi'\left(\rho_{II}(\xi_{II} - d)\right) \end{bmatrix}^{-1} G\exp\left(ik_{III}d\right) \qquad 5.55$$

Equation 5.55 can be substituted in 5.54 resulting in an equation of the form

$$\begin{bmatrix} A \\ B \end{bmatrix} = G\exp\left(ik_{III}d\right)\begin{bmatrix} C \\ D \end{bmatrix} \qquad 5.56$$

The transmission coefficient

$$T = \frac{k_{III}}{k_I}|t|^2 \qquad 5.57$$

where

$$t = \frac{G}{A} \qquad 5.58$$

can be obtained from equation 5.56. We normally define a transmission matrix S which is the matrix that relates output to input. If we had allowed an incoming wave coming from the right with coefficient H in layer III, which would have given also a current from right to left, the (2x2) transmission matrix S would obey an equation of the form

$$\begin{bmatrix} G \\ H \end{bmatrix} = \begin{bmatrix} S \end{bmatrix} \begin{bmatrix} A \\ B \end{bmatrix}$$

5.59

As noted previously and as should be apparent now, this technique is only useful for simple barriers. For barriers of a general shape, one has to revert to the WKB approximation discussed in an introductory way in chapter 1 and more thoroughly in the next chapter.

5.6 THE RESONANT TUNNELLING DIODE OR RTD

The PN and Schottky diodes that we have discussed in chapter 3 and all the transistors we will discuss are based mainly on the drift–diffusion mechanism. The RTD, as the name signifies, is based on the tunnelling mechanism, and it cannot be analyzed by any classical method but only by such methods as the Landauer approach, so it is reasonable that we present it here.

The RTD involves the tunnelling of electrons through two barriers separated by a quantum well. A physical realization of such a structure is accomplished by the 5-layer sandwich of $GaAs/Al_xGa_{1-x}As/GaAs/Al_xGa_{1-x}As/GaAs$. The two outer layers of GaAs are usually overdoped so as to act as contacts-leads, see figure 5.11. The compound $Al_xGa_{1-x}As$ is an alloy of GaAs and AlAs. The latter has a band-gap of $2.13eV$, whereas GaAs has a band-gap of $1.42eV$. Therefore the conduction band offset $\Delta E_C(x) \approx 60\% \Delta E_g(x)$ acts as a barrier to flow from GaAs to $Al_xGa_{1-x}As$. The stoichiometric index x varies between 0 and 0.43. The rather amazing feature of such a structure is that it exhibits negative differential resistance i.e. $dI/dV < 0$. This pair of semiconductors is not the only pair that exhibits resonant

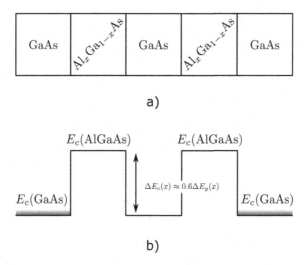

a)

b)

FIGURE 5.11   (a) A five-layer sandwich of alternating $GaAs$ and $Al_xGa_{1-x}As$ layers which constitutes an RTD diode, and (b) the conduction band edge E_C of this system.

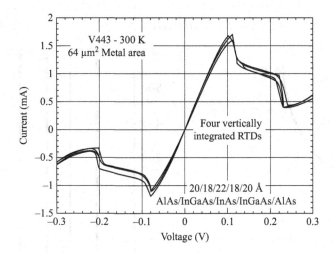

FIGURE 5.12 The experimental $I-V$ characteristic of a III–IV system showing negative differential resistance. Figure reproduced from T. S. Moise et al J. Appl. Phys. 78, 6305 (1995).

tunnelling. Many pairs of III–IV and II–VI semiconductors exhibit resonant tunnelling. Figure 5.12 shows the experimental current-voltage characteristic of a III–V resonant tunnelling diode and one can clearly see regions in the characteristics where the current is decreasing while the voltage is increasing. A pair of II–VI semiconductors that exhibits resonant tunnelling is ZnMgO. The band diagram of any such structure is shown in figure 5.13. Layers II and IV constitute the barriers while layer III is the quantum well.

In the previous section, we have developed all the methodologies needed to analyze such a structure. In the two outer layers where the bands are flat, the eigenfunctions will be planewaves with an incoming plane wave and a reflected one in the left layer and only a transmitted one in the right one. On the other hand, in the layers where the potentials (or

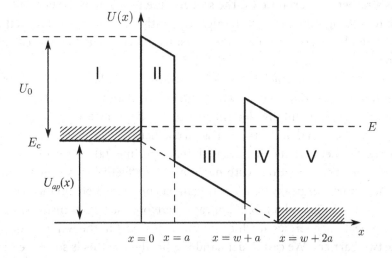

FIGURE 5.13 The band structure diagram of the system in figure 5.11 under the application of a voltage difference.

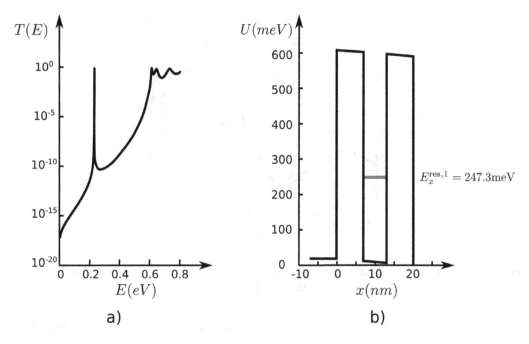

FIGURE 5.14 Calculated transmission coefficient and experimentally determined quasi-bound eigenstate for the RTD with $ZnO/Zn_{1-x}MgO$ alternating layers (calculation by the author).

conduction band edges) vary linearly with length, the wavefunctions will be combinations of first and second kind Airy functions as described in detail in the previous section. Then applying the continuity conditions for the wavefunction Ψ and its derivative $\dfrac{1}{m^*}\dfrac{d\Psi}{dx}$ at the interfaces, we get a matrix equation relating the coefficients of the wavefunctions. We leave this as a rather long exercise for the student (see problem at the end of the chapter). The procedure is straightforward but the algebra is rather tedious. If such a methodology is applied to this double barrier problem, the transmission coefficient can be calculated with the mathematical apparatus given in section 5.5.

The transmission coefficient $T(E)$ thus calculated for the system $ZnO/Zn_{1-x}Mg_xO/ZnO/Zn_{1-x}Mg_xO/ZnO$ with $x = 0.2$ and $\Delta E_C = 0.6 eV$ is shown in figure 5.14a. The experimentally determined band diagram for this case is shown in figure 5.14b. It can be seen that the transmission coefficient has a sharp maximum, equal to one, at roughly the energy that coincides with the experimental energy of a bound state in the ZnO well. Further calculations with much heavier effective masses for both materials but smaller ΔE_C show four peaks with $T = 1$, again at energies where bound states appear in the well, see figures 5.15a and 5.15b. It appears therefore that the transmission coefficient has very sharp peaks at energies at which bound states exist in the well that is sandwiched between the two barriers. We can understand physically why this is so. Since the well is of a nanometric length, it can accommodate a small number of discreet states. On these states an electron can "momentarily step on" or more precisely resonate with.

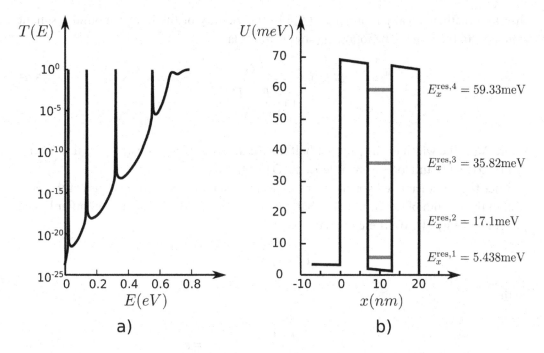

FIGURE 5.15   Same as in figure 5.14 but with different material parameters; see text.

For a more accurate description of the currents in an RTD we need a 3-dimensional (3D) model. In the left lead, where bands are flat, we can write for the 3D wavefunctions

$$\Psi^L(r_\perp, x) = A e^{i k_\perp \cdot r_\perp} e^{i k_x x} \tag{5.60}$$

where x denotes the direction along the RTD and $r_\perp$ denotes the 2-dimensional position vector perpendicular to x. k_x and $k_\perp$ are the wavevectors along the respective directions. In the right lead of the RTD, where again the bands are flat, we can write for the wavefunction:

$$\Psi^R(r_\perp, x) = G e^{i k_\perp \cdot r_\perp} e^{i k_x x} \tag{5.61}$$

The potential energy in the effective mass equation is only a function of x, i.e. $U = U(x)$. Hence this equation can be separated in longitudinal (x) and perpendicular ($r_\perp$) components. The longitudinal x component will be identical to what we have described in the previous section 5.5, so that we can write

$$T(E_x) = \left| \frac{G(E_x)}{A(E_x)} \right|^2 \tag{5.62}$$

As already shown, $T(E_x)$ has sharp peaks at the energies at which quasi-bound states appear in the well. (The prefix "quasi" is used because electrons do not stay indefinitely there but only for a limited time on their transport from one electrode to the other). We can therefore mathematically represent $T(E_x)$ in the vicinity of the ith quasibound-resonant state $E_x^{res,i}$ in the well by the following simple formula

$$T(E_x) = \frac{Q}{1 + \left(\dfrac{E_x - E_x^{res,i}}{\Delta E}\right)^2} \qquad \text{5.63}$$

where ΔE is the width of the peak (at half maximum) and Q is a constant equal to 1 for a symmetric RTD and for an asymmetric RTD $Q < 1$.

Since there is an applied voltage V at the electrodes of the RTD, the energies of the electron in the bottom of the conduction band in the left and right electrode will differ by the potential energy eV. We therefore have

$$E_x^L - E_x^R = eV \qquad \text{5.64}$$

so that

$$E_x^L - E_x^R = \frac{\hbar^2 \left(k_x^l\right)^2}{2m^*} - \frac{\hbar^2 \left(k_x^r\right)^2}{2m^*} = eV \qquad \text{5.65}$$

where L,R stand for the left and right reservoir or lead. We initially assume as in section 5.3 and section 5.5 that the conduction edge E_C^R is lowered by eV and E_C^L is unchanged. To use the Landauer formula we must now calculate N_L and N_R of equation 5.33, which are usually called supply functions. They are the number of electrons in the left and right lead reservoir respectively that have the energy E_x, the one parallel to the channel, fixed, i.e. they are the number of electrons in the leads incident on the barriers with a given normal to the barrier energy. Note that E_x, being along the channel, is also normal to the barrier. These functions can be calculated by employing our result for the density of states, DOS, of a 2-dimensional system. The leads of the RTD are not necessarily 2-dimensional systems since they are not necessarily of nanometric length, but the density of states of a 3-dimensional material with the energy in 1 dimension fixed, is mathematically the same as that of the DOS of a 2-dimensional system with a single energy in the remaining (confined) dimension.

Let the cross-section of the device be S. Then the required numbers in the leads are, using equation 5.12 with one term only in the summation and omitting the superscripts L, R for the moment,

$$N = \int_0^\infty S g^{2D}(E) f(E, E_F) dE =$$

$$= \int_0^\infty \frac{Sm^*}{\pi\hbar^2} \frac{\Theta(E_x - E_F)}{1 + exp\left(\dfrac{E - E_F}{KT}\right)} dE =$$

$$= \frac{Sm^*}{\pi\hbar^2} \int_{E_x}^{\infty} \frac{dE}{1+exp\left(\frac{E-E_F}{KT}\right)}$$

5.66

where in 5.66 f stands for the Fermi–Dirac function (written more explicitly) and K is Boltzmann's constant. The above integral can be put into the form $\int dx/1+x$ so that N_L and N_R become (after a few manipulations)

$$N_L = \frac{SKTm^*}{\pi\hbar^2} ln\left(1+exp\left(\frac{E_F^L - E_x}{KT}\right)\right)$$

5.67a

$$N_R = \frac{SKTm^*}{\pi\hbar^2} ln\left(1+exp\left(\frac{E_F^R - E_x}{KT}\right)\right)$$

5.67b

Application of Landauer's formula, equation 5.33 gives for the current density $J = I/S$

$$J = \frac{em^*kT}{2\pi^2\hbar^3} \int_{E_C^{em}}^{\infty} T(E_x) ln\left[\frac{1+exp\left(\frac{E_F^L - E_x}{KT}\right)}{1+exp\left(\frac{E_F^L - E_x - eV}{KT}\right)}\right] dE_x$$

5.68

In using equation 5.68, care should be exercised in avoiding regions of the energies E_x at which the initial and final state of the electron are both occupied as shown in figure 5.16 and hence not contributing to the integral in 5.68. Such a case does not arise if $eV = E_F^R - E_F^L \gg KT$.

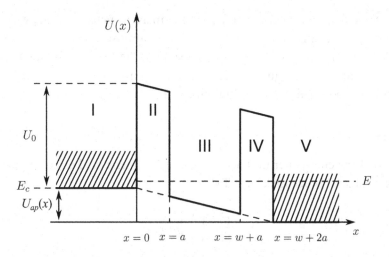

FIGURE 5.16 An arrangement of the bands in which transmission is not allowed because an electron goes from a filled state to a filled state also.

In fact if $eV \gg KT$ only the current $L \rightarrow R$ contributes to the total current. Assuming that the edge of the left conduction band is lifted by eV, compared to the unperturbed Fermi level and the right one remains unchanged (as shown in figure 5.16) and writing $E_F^L = E_F^R = E_F^0$ at $V = 0$ equation 5.68 becomes for large V

$$J = \frac{em^* kT}{2\pi^2 \hbar^3} \int_{E_C^{em}}^{\infty} T(E_x) \ln\left[1 + \exp\left(\frac{E_F^0 + eV - E_x}{KT}\right)\right] dE_x \qquad 5.69$$

The limit of $T \rightarrow 0$ (low temperatures) allows the following simplification

$$N_L(E_x) = \frac{Sm^*}{\pi \hbar^2}\left(E_F^L - E_x\right) \qquad 5.70$$

Then equation 5.69 transforms into

$$j = \frac{em^*}{2\pi^2 \hbar^3} \int_{E_C^L}^{E_F^L} \left(E_F^L - E_x\right) T(E_x) dE_x, \text{ for } eV > E_F^R - E_C^R \qquad 5.71$$

where $E_F^L = E_F^0 + eV$.

If the Lorentzian in equation 5.63 is sharp around any of the resonances $E_x^{res,i}$ the integration in equation 5.71 can be performed analytically: any such integration gives the value at $E_x^{res,i}$ of the remaining part of the integrand times the width of the Lorentzian ΔE. Hence we get for the current density near a maximum

$$j = \frac{em^* \Delta E}{2\pi^2 \hbar^3}\left(E_F^0 + eV - E_x^{res,i}\right) Q \qquad 5.72$$

where $Q = 1$ for a symmetric barrier and $Q < 1$ for an asymmetric one. So, very near the ith maximum at $E_x^{res,i}$ the current density first increases almost linearly as we approach $E_x^{res,i}$ and then decreases linearly and then drops abruptly due to the Lorentzian shape of the transmission coefficient. At higher temperatures the approximation 5.70 is not valid so that a smooth curve is obtained as shown in figure 5.12.

PROBLEMS

5.1 Prove that the energy per unit area U/A in a 2-dimensional electron gas, when only the first sublevel is occupied is at T=0

$$\frac{U}{A} = \left(\frac{1}{2}\right) n E_F$$

where n is the number of electrons per unit area.

5.2 Use the calculated wavefunctions of problem 1.2 to express the transmission over a potential step in matrix form as $\begin{matrix} C \\ D \end{matrix} = T \begin{matrix} A \\ B \end{matrix}$ where A,B,C,D are the coefficients of the incident and reflected waves from either side of the step and T is a matrix.

5.3 Show that the supply function $(N_L - N_R)$ of the resonant tunneling diode (equation 5.67) is proportional to V for small voltages V.

5.4 Given the band-structure diagram of figure 5.13, write down the wavefunctions in each of the layers I to V. Follow the methodology of section 5.5. Then write down the conditions for continuity of the wavefunction and its derivatives at each of the interfaces of the system.

5.5 Starting from the conditions for the continuity of the wavefunction and its derivatives at the interfaces derived above, use the methodology of section 5.5 to write the outcoming wave in matrix form in terms of the incident and reflected waves.

III

Devices

CHAPTER 6

Field Emission and Vacuum Devices

6.1 INTRODUCTION

Vacuum or field emission devices where electrons are extracted from metals or semiconductors are a minority in the world of devices. However, we will begin our discussion of devices with field emission ones because of their simplicity: they involve only two electrodes, usually called anode and cathode, between which conduction by tunnelling through vacuum occurs. Transistors, on the other hand, which are made of semiconducting materials, are three terminal devices in which the third electrode, called gate, modifies the conductance between the other two. The above statement is incorrect from a historical point of view. The vacuum valve was the predecessor of the modern transistor and was the first tool to produce electronic amplifiers, oscillators, etc. Actually, the vacuum valve in solid state form—a transistor with vacuum as its channel—seems to have been revived lately but our choice of starting from vacuum devices, mainly diodes, stems from the simplicity of not only their configuration but also of applying our quantum formalism: the current is only from the cathode to the anode and there is no current in the opposite direction as there is in all semiconductor devices. A further simplification is that collisions during emission do not play a significant role and are usually omitted. We have emphasized many times that due to the nanometric size of present day devices, only a quantum approach to conduction can yield accurate results, such as the Landauer theory of conduction which relates the conductive properties to the transmissive properties of a medium. In the previous chapter, we have shown how to calculate the transmission coefficient of mathematically simple barriers with no collisions. Unfortunately, such simple barriers rarely occur in modern diodes and transistors. An exception is the RTD discussed in the previous chapter where the vast majority of the current is a tunnelling current. In MOSFETs and other forms of transistors, the conduction is 3-dimensional and over the barrier, not through the barrier. Therefore, a proper knowledge of the wavefunctions themselves is mandatory, not just of the transmission coefficient, as will become evident in the next chapter. A method capable of doing this for arbitrary but slowly varying potentials without resort to numerical calculations is the WKB[1] method which we initially touched upon in chapter 1. We now discuss it at some depth as it will be useful in many cases of device analysis.

6.2 THE 1-DIMENSIONAL WKB EQUATION

Although the name of the approximation bears the initials of three authors that published, each separately, a corresponding paper (all in 1926) in the field of Quantum Mechanics, the history of the WKB approximation goes further back in time. Most of the essential ideas and methods had been worked out by Lord Rayleigh as far back as 1912, if not earlier in 1896, with the publication of his book "Theory of Sound". As the name of the book implies, the mathematical method was aimed at the physical problem of sound propagation. Some authors refer to this method also as the JWKB approximation after an earlier publication (in 1923) by H. Jeffreys. Before tackling the 1-dimensional case in depth, we devote a few lines to show the connection between the classical wave equation and that of Schroedinger.

Lord Rayleigh tackled the problem of a scalar wave Φ propagating in a slowly varying medium characterized by a refractive index n(r). This is equivalent to solving the wave equation

$$\nabla^2 \Phi + k^2 n^2 \Phi = 0 \qquad 6.1$$

If the refractive index $n(r)$, which is slowly varying, were a constant equal to n_0, the solutions of equation 6.1 would be plane waves of the form

$$\Phi \propto e^{in_0 k \cdot r} \qquad 6.2$$

It made sense therefore to try a solution in at least the 1-dimensional but position dependent case of the form

$$\Phi \propto e^{ik\sigma(r)} \qquad 6.3$$

where $\sigma(r)$ is a slowly varying function of r.

The Schroedinger equation can be put into the form of 6.1 by simply writing it as

$$\left[\nabla^2 + \frac{2m}{\hbar^2} (E - V(r)) \right] \Psi = 0 \qquad 6.4$$

Then the $\frac{2m}{\hbar^2}(E - V)$ term above is the equivalent of $k^2 n^2$. Alternatively we may divide and multiply the second term in 6.4 by the energy E to get

$$\left[\nabla^2 + \frac{2mE}{\hbar^2} \left(1 - \frac{V(r)}{E} \right) \right] \psi(r) = 0$$

---

[1] From the names of the physicist Wentzel, Krammer, and Brillouin, who developed it in the early 1900's.

But

$$\frac{2mE}{\hbar^2} = (p/\hbar)^2 = k^2 = \left(\frac{2\pi}{\lambda}\right)^2$$

where p = momentum and λ is wavelength. Hence equation 6.4 may be put again into the form of 6.1 if we assign

$$n(r) = \sqrt{1 - \frac{V(r)}{E}}$$ 6.5a

and

$$k^2 = \frac{2mE}{\hbar^2}$$ 6.5b

The analogy between the Schroedinger equation and the classical wave equation does not seem at first sight to be universal because in the case of the Schroedinger equation the term $E - V$ may be either negative or positive, whereas the term $k^2 n^2$ is positive for wave propagation in vacuum. We remind the reader, however, that metals have an almost imaginary refractive index so that n^2 in the case of metals is negative and the analogy is complete. It is worthwhile noting that in both types of equations the solutions are travelling waves if the quantity $\frac{2m}{\hbar^2}(E - V)$, or equivalently kn^2, is positive or decaying functions of space if the quantity $(E - V)$ is negative. Figures 6.1a and 6.1b portray these two cases.

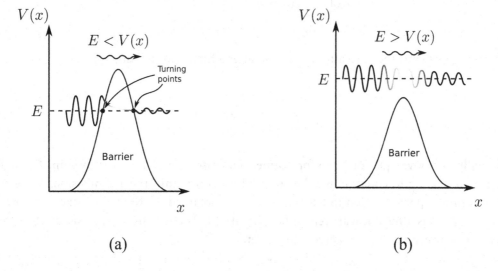

(a)   (b)

FIGURE 6.1   Transmission of a wavefunction (a) through a potential barrier and (b) over a potential barrier.

We now return in detail to the 1-dimensional case. We mentioned in the introduction to this section that the WKB method is valid for slowly varying potentials. We need to define what is meant by "slowly varying" more accurately. An obvious mathematical expression of this statement would be

$$\left|\frac{1}{n}\frac{dn}{dx}\right| \ll \frac{1}{\lambda}$$

6.6

where λ is the de Broglie wavelength of electrons. Let the average of the absolute quantity in 6.6 be denoted by

$$\overline{\left|\frac{1}{n}\frac{dn}{dx}\right|} = \frac{1}{L}$$

6.7

Then a better expression than 6.6 will be

$$\frac{1}{L} \ll \frac{1}{\lambda} \Rightarrow L \gg \lambda$$

Substituting equation 6.3 in 6.1 we have in 1 dimension by successive differentiation

$$\frac{\partial^2}{\partial x^2} e^{ik\sigma(x)} + k^2 n^2 e^{ik\sigma(x)} = 0 \Rightarrow$$

$$\Rightarrow \frac{\partial}{\partial x}\left[e^{ik\sigma(x)}\left(ik\frac{\partial\sigma(x)}{\partial x} \right)\right] + k^2 n^2 e^{ik\sigma(x)} = 0$$

Denoting by prime the derivative $\frac{\partial}{\partial x}$ we get after a further differentiation

$$\frac{i}{k}\sigma'' - (\sigma')^2 + n^2 = 0$$

6.8

Actually the above equation holds for 3 dimensions also, as one can verify by going through the steps leading to equation 6.8, but we will remain within the 1-dimensional case. A zeroth order approximation to 6.8 can be readily obtained: if we assume that the slowly varying $\sigma(x)$ is almost linear (remember that if $n(x)$ is constant $\sigma(x)$ is just x) then the second derivative σ'' is nearly zero and we get

$$\left(\sigma_0'\right)^2 = n^2 \Rightarrow \sigma_0' = \pm n$$

6.9

Integrating

$$\sigma = C \pm \int n(x)\,dx \qquad\qquad 6.10$$

where C is the constant of integration.

Now assume first that $n(x) > 0$ $\big($or $E > V(x)\big)$ so that the square root above is real. The opposite inequality will be examined next. We can get a better approximation than equation 6.9 as follows: we expand σ in powers of

$$\lambda/2\pi L \equiv a \ll 1 \qquad\qquad 6.11$$

Note that we retain the factor 2π because $k = 2\pi/\lambda$. We have

$$\sigma = \sigma_0 + a\sigma_1 + a^2\sigma_2 + \cdots \qquad\qquad 6.12$$

If we keep the first two terms now, as a better approximation to equation 6.9, and substitute these terms into equation 6.8 we get

$$\frac{i}{k}\left(\sigma_0 + a\sigma_1\right)'' - \left(\sigma_0' + a\sigma_1'\right)^2 + n^2 = 0 \Rightarrow$$

$$\Rightarrow \frac{i}{k}\left(\sigma_0'' + a\sigma_1''\right) - \sigma_0'^2 - 2a\sigma_0'\sigma_1' - a^2\left(\sigma_1'\right)^2 + n^2 = 0$$

Great care should be exercised in approximating the above equation. The term $a^2\left(\sigma_1'\right)^2$ can be dropped as it is to second order in the parameter a. A further step is to note that $\sigma_0'^2$ and n^2 cancel out if use of equation 6.9 is made. The last step in our approximations is to neglect σ_1'', just as in our zeroth order approximation we neglected σ_0''. But in this order the second derivative σ_0'' should be kept. We then get

$$\frac{i}{k}\sigma_0'' - 2a\sigma_0'\sigma_1' = 0 \Rightarrow$$

$$\Rightarrow \sigma_1' = \frac{i}{2ka}\frac{\sigma_0''}{\sigma_0'} = i\frac{L}{2}\frac{\sigma_0''}{\sigma_0'} \Rightarrow$$

In the equation above we have used equation 6.11, so that

$$\sigma_1 = \frac{iL}{2}\ln\left(\sigma_0'\right) + \text{a constant} \qquad\qquad 6.13$$

Substituting for σ_0 from equation 6.10 and for $a\sigma_1$ from 6.13 into equation 6.12 and keeping the first two terms we have

$$\sigma(x) = C \pm \int n(x)dx - \frac{1}{2}ln[n(x)]$$

6.14

where the constant in 6.13 has been absorbed in the constant C. Substituting 6.14 in our basic equation 6.3 we get

$$\Phi(x) = \frac{C}{\sqrt{n(x)}} exp\left[\pm i \int kn(x)dx\right]$$

6.15

Now we multiply and divide the pre-exponential factor by $\sqrt{k}$ and absorb the latter in the numerator into C. Then using equations 6.5a and 6.5b to transform back to the domain of Quantum Mechanics (where now the field is $\Psi(x)$) we have

$$\Psi(x) = \frac{G}{\sqrt{\frac{2m(E-V)}{\hbar}}} exp\left[\pm i \int \frac{\sqrt{2m(E-V)}}{\hbar}dx\right]$$

6.16

In equation 6.16 both solutions with the (+) or (−) sign are acceptable solutions, so the general solution may be written as

$$\Psi(x) = \frac{A}{\sqrt{\frac{2m(E-V)}{\hbar}}} exp\left[+\frac{i\sqrt{2m(E-V)}}{\hbar}dx\right] + \frac{B}{\sqrt{\frac{2m(E-V)}{\hbar}}} exp\left[-\frac{i\sqrt{2m(E-V)}}{\hbar}dx\right]$$

6.17

What happens if $E < V(x)$ or equivalently $kn^2 < 0$? Then the solutions are real exponentials and not imaginary exponentials, i.e. travelling waves. They can be written in the form

$$\Psi(x) \propto e^{k\sigma(r)}$$

6.18

An exactly parallel analysis to what we have presented so far, leads to

$$\Psi(x) = \frac{C}{\sqrt{\frac{2m(V-E)}{\hbar}}} exp\left[\frac{\sqrt{2m(V-E)}dx}{\hbar}\right] + \frac{D}{\sqrt{\frac{2m(V-E)}{\hbar}}} exp\left[-\frac{\sqrt{2m(V-E)}dx}{\hbar}\right]$$

6.19

Both equations 6.17 and 6.19 present a mathematical problem at points x such that $V = E$, i.e. where the potential energy is equal to the total energy. Then the kinetic energy is zero and hence the classical velocity is zero. From both equations 6.17 and 6.19 we deduce that the wavefunction Ψ goes to infinity. In this case the WKB approximation breaks down. The points in space at which this happens are shown in figure 6.1a and are designated "turning points" because classically the velocity would have changed direction at these points. These points are crucial because it is at these points we have to calculate the wavefunction to obtain the transmission of a wave through a barrier such as the one shown in figure 6.1a. The transmission coefficient defined by

$$T = \left[\frac{\Psi(\text{transmitted})}{\Psi(\text{incident})} \right]^2 \times \frac{v(\text{transmitted})}{v(\text{incident})}$$

is divergent at these points. In this particular case one has to expand the potential around the turning points x_1 and x_2 as $V(x) = V(x_i) + \beta_i x$ and seek the solutions of the Schroedinger equation in terms of Airy functions. The outcome of such a procedure is the transmission coefficient given by the formula we have already presented in chapter 1. We have given a simplified proof there. The more rigorous proof is quite lengthy with no further physical insight so we omit it. For completeness we repeat the equation for the transmission coefficient of chapter 1 here (equation 1.65)

$$T(E) = exp\left[-2\int_{x_1}^{x_2} \frac{\sqrt{2m(V(x) - E_x)}}{\hbar} dx \right] \qquad 6.20$$

where in 6.20 we have written explicitly E_x to denote the energy along the x-direction for future use, when more dimensions will be involved.

This completes our exposition of the WKB method. Although we had presented the above formula in chapter 1 it was necessary to go into more detail in order to obtain the wavefunctions as these will be necessary in our quantum description of nanoscale transistors.

6.3 FIELD EMISSION FROM PLANAR SURFACES

All objects being finite end up with a surface facing air or near vacuum. We have not discussed surfaces so far but they are not very different from the interfaces we have discussed. Consider a finite 1-dimensional array of atoms as in figure 6.2. Every electron in the bulk of the solid (i.e. in the array) experiences electrostatic interactions from the electrons of the neighbouring atoms from both left and right giving rise to the crystalline potential shown in the figure. In chapter 2 we considered the crystalline potential to be infinitely periodic. But in the case of a finite crystal, simplistically portrayed in figure 6.2, the end atom on the far right has no atom to its right and therefore its electron sees no electrostatic push from the right. Therefore, its electrons do not experience the potential seen by other electrons in the bulk but an increased potential in the form of a surface barrier as shown in figure 6.2.

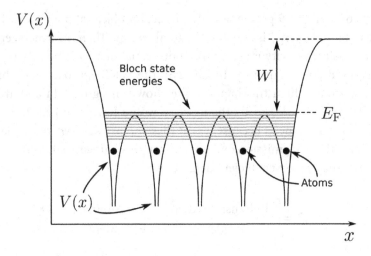

FIGURE 6.2 Schematic form of the crystalline potential in the bulk of a solid and near its surfaces. Near the surface, a potential barrier is created. Also shown are the Bloch state energies and the Fermi level.

The same argument holds for the electrons of the last atom on the far left. The minimum energy needed to take an electron out of the solid is the energy needed to overcome this surface barrier and is called workfunction. It is denoted by W in figure 6.2. We initially introduce briefly this barrier in connection with the Schottky diode in chapter 3. This barrier is similar in nature to the barrier between a metal and a semiconductor in the corresponding interface in the sense that it derives from an asymmetry of the nearby atoms. The above description is not complete. Since the electrons in the surface atoms no longer see a symmetric potential, they will tend to spill out towards the vacuum side leaving a positive charge behind and thus creating a dipole layer near the surface. This is shown in figure 6.3.

It is customary to simplify the above picture by using the Sommerfield model. Then for metals we only need to show the conduction band, the corresponding Fermi level, and the surface barrier we have just discussed. The energy band diagram which takes into

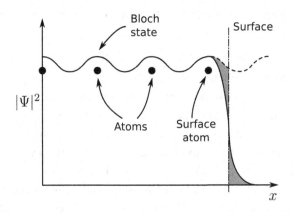

FIGURE 6.3 A dipole layer is created near the surface of a solid because the Bloch eigenstates "spill out" of the solid surface.

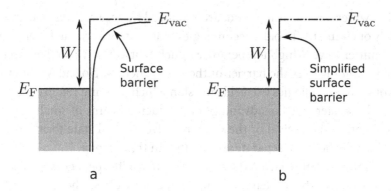

FIGURE 6.4 The actual surface barrier shown in (a) may be approximated as having zero width as shown in (b).

account this simplification is shown in figure 6.4a. Usually the surface barrier is further approximated by a vertical line extending up to the vacuum level. Then the energy band diagram of figure 6.4a is further simplified into the band diagram of figure 6.4b. Now let us assume a parallel plate capacitor geometry where two parallel metal plates with a distance d between them, are connected by a voltage V, giving rise to a constant electric field $\mathcal{E} = V/d$ and an electrostatic energy $V_{el} = -e\mathcal{E}x$ in the space between the plates. The energy diagram corresponding to this situation with the surface barrier simplified as in 6.4b is shown in figure 6.5. Assume that temperature is low and the electric field $\mathcal{E}$ is in the range of a few V/nm. Then tunnelling occurs as shown by the arrow and electrons are transported from the left to the right metal plate. The phenomenon is called Field (assisted electron) Emission.

Field Emission has many applications notably in Microscopy, Lithography, and microwave production and amplification. In all these cases a beam of electrons is the necessary input and this is provided by Field Emission. In Microscopy, the electron beam (presently of nanometric width) is used to probe the structure of a specimen, whereas in Lithography the beam is used to "carve" characteristics on semiconductors which transform them into

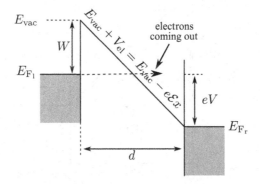

FIGURE 6.5 The barrier for emission under the action of an electric field $\mathcal{E}$ is triangular in the approximation of zero width barrier.

devices. Microwaves, on the other hand, are produced by accelerating a beam of electrons either linearly or circularly. In older technologies that beam was created thermionically, i.e. heating the emitter to very high temperatures (such that $k_B T \approx W$) so that electrons could be emitted by climbing over the barrier of the workfunction. Nowadays, however, a large proportion of such technologies use field emission which does not require a heating circuit and, as we will see later, has the advantage of producing beams of small angles. Last but not least we mention the revival of the vacuum valve in solid state form called "vacuum transistor". We will allocate a separate section to it in this chapter.

When an electron is emitted from the surface of a metal it experiences the force due to the surface barrier potential which extends only up to roughly 2Å from the surface. Although this potential energy is created by the complicated electron-electron interactions at the surface a very good approximation can be obtained if the image potential of electrostatics is used for it. Therefore, the potential energy that an electron will see on its way to vacuum, called tunnelling potential for short V_{tun}, is (choosing the x direction as the one perpendicular to the surface)

$$V_{tun}(x) = E_F + W - e\mathcal{E}x - e^2/16\pi\varepsilon_0 x =$$

$$= E_{vac} - e\mathcal{E}x - \frac{e^2}{16\pi\varepsilon_0 x} \qquad 6.21$$

where the $(-)$ sign of the third term of the first line of 6.21 comes from the charge of the electron and that of the fourth term from the fact that the interaction is attractive. The graph of $V_{tun}(x)$ is shown in figure 6.6.

The above potential energy will enter the effective mass equation which will result in the x direction after the separation of variables. In particular, if the envelop function is assumed to be of the form

$$\Phi_{k_\parallel}(\eta_\parallel, x) = e^{ik_\parallel \cdot \eta_\parallel} \mathcal{X}(x) \qquad 6.22$$

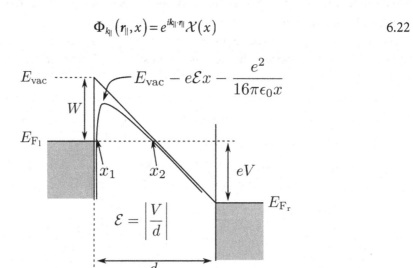

FIGURE 6.6 The potential barrier for emission when the image potential is used to represent the surface.

with $k_{\parallel}$ denoting the wavevector parallel to the surface we will get after separation of variables in the x direction

$$\left[-\frac{\hbar^2}{2m^*}\frac{d^2\mathcal{X}(x)}{dx^2}+V_{tun}(x)\right]\mathcal{X}(x)=E_x\mathcal{X}(x) \qquad 6.23$$

with $E_x = E - \hbar^2 k_{\parallel}^2/2m$, the energy along x. The current that comes out of the barrier $V_{tun}(x)$ can now be calculated in the same manner that the current from a resonant tunnelling diode was calculated. In fact, equation 5.69 is directly applicable here because the applied potential may be considered large enough, so that only a Left to Right current exists. Rewriting equation 5.69 we have

$$J = \frac{em^*KT}{2\pi^2\hbar^3}\int_{E_b}^{\infty} \ln\left(1+exp\left(\frac{E_F - E_x}{kT}\right)\right)T(E_x)dE_x \qquad 6.24$$

where we have replaced the bottom of the conduction band of an RTD emitter E_C^{em} by the bottom of the conduction band of the metal E_b and have written E_F for $E_F^0 + eV$. Energies are measured from the bottom of the band (so we can set $E_b = 0$).

Certain points should be made clear. Just as in the case of the RTD (section 5.6), the bottom of the conduction band E_b is assumed to be flat, i.e. there is no variation with x inside the metal. This is due to the fact that the electric field $\mathcal{E}$ terminates on the metal surface. (Actually electric fields penetrate metals by approximately half a monolayer, but this is always neglected). On the other hand, electric fields penetrate semiconductors substantially (i.e. by nanometers) and hence the ln term, called the supply function, has to be evaluated for semiconductors right at the surface. A second point is that, due to the complexity of the barrier, the transmission coefficient $T(E_x)$ can not be evaluated exactly and hence the WKB approximation has to be used, i.e. equation 6.20.

Now an inspection of figure 6.5 and an examination of equation 6.20 show that $T(E_x)$ is a rapidly decaying function as E_x decreases. This is easily seen if we write compactly

$$T(E_x) = exp\left[-G(E_x)\right] \qquad 6.25$$

where $G(E_x)$ is called the Gamow exponent and is equal to

$$G(E_x) = 2\int_{x_1}^{x_2} \frac{1}{\hbar}\sqrt{2m(E_{vac} - e\mathcal{E}x - e^2/16\pi\varepsilon_0 - E_x)}dx \qquad 6.25a$$

The above equation is the outcome of substituting equation 6.21 in the expression for the transmission coefficient, equation 6.20, and of using the vacuum mass of the electron instead of the effective mass. Note that x_1 does not coincide with the emitting surface but is slightly above it, As E_x decreases, i.e. moves away from either E_F or E_{vac}, the above integral increases and hence $T(E_x)$ rapidly decreases. It makes sense therefore to expand $G(E_x)$ in a Taylor expansion and keep the first two terms only. The expansion is performed around the energy point $E_x = E_F$, i.e. where the normal to the barrier component of energy is equal to E_F. With $\Delta E_x = E_x - E_F$, we have

$$G(E_x) = G(E_x = E_F) + \Delta E_x \left. \frac{\partial G(E_x)}{\partial E_x} \right|_{E_x = E_F} =$$

$$= G(E_F) + \Delta E_x d_F \qquad 6.26$$

where

$$d_F = \left. \frac{\partial G(E_x)}{\partial E_x} \right|_{E_x = E_F} \qquad 6.26a$$

If equation 6.26 is substituted into equation 6.24 we get a very good approximation for the current density J

$$J = \frac{emKT}{2\pi^2 \hbar^3} T(E_F) \int_{E_b}^{\infty} exp(d_F(E_x - E_F)) ln\left[1 + exp\left(\frac{E_F - E_x}{KT} \right) \right] dE_x \qquad 6.27$$

One should not confuse the thermal energy KT with the transmission at the Fermi energy $T(E_F)$. The integral in 6.27 can be evaluated analytically. The steps are purely mathematical so we relegate the proof to Appendix E. The result is

$$J = \frac{emKT \cdot T(E_F)}{2\pi d_F \hbar^3 sin(\pi d_F KT)} \qquad 6.28$$

As can be seen from 6.28, d_F has the dimensions of $(energy)^{-1}$ and denotes the energy distance one has to go below E_F so that $T(E)$ becomes negligible. It is usually of the order of $1eV$. On the other hand kT at room temperature is approximately $1/40eV$ so that the term $d_F kT$ is much smaller than 1. We may expand to second order the term $1/sin(\pi d_F kT)$ to get finally

$$J = \frac{emT(E_F)}{2\pi^2 \hbar^3 (d_F)^2} \left[1 + \frac{(\pi d_F KT)^2}{6} \right] \qquad 6.29$$

It is worthwhile pointing that only a small portion of the whole $\mathbf{k}$-space contributes to emission. This portion is shown in figure 6.7, where we represent the $E(\mathbf{k})$ space in contracted 2 dimensions, one for the dimension E_x, the normal energy, and the other for the energy $E_{\parallel}$, the energy parallel to the barrier. Since the transmission coefficient is a function

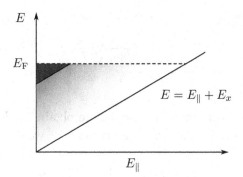

FIGURE 6.7   In a phase diagram with the parallel to the barrier energy on the x-axis and the total energy on the y-axis most of the field-emitted current comes from the small shaded upper triangle.

of E_x only, any increase $E_{\parallel}$, which corresponds to a decrease of E_x, will also lower the transmission coefficient. Hence emission comes from the states which form the shaded triangle in figure 6.7. It is customary to call the expression in the brackets in equation 6.29 as $\Theta(T)$ so that we may write

$$J = \frac{emT(E_F,0)}{2\pi^2\hbar^3(d_F)^2}\Theta(T) \qquad 6.30$$

where $T(E_F,0)$ is the transmission coefficient at the Fermi energy at zero temperature. Obviously the expression before the factor $\Theta(T)$ is the zero temperature current density.

All that remains now is to calculate $T(E_F)$, the transmission coefficient at the Fermi energy. Once the Gamow exponent is known, $T(E_F)$ immediately follows. This is given by equation 6.25a. It is customary to perform the calculation in two stages: we first ignore the image potential. This is actually what has happened historically—the researchers of the 1930s tried the triangular barrier as a first approximation and recognized that this choice was inadequate because more than the calculated current was coming out of their emitters. We will only follow this historical path because it leads us smoothly to a complicated result.

If the image potential is neglected, the integral in 6.25a is easily calculated since the integrand is of the form $\sqrt{a-bx}$ and the limits of integration are $x_1=0$ and $x_2=(E_{vac}-E_x)/(e\mathcal{E})$. Denoting by the superscript (tr), the results for the simple triangular barrier we get by a direct calculation

$$G^{tr}(E_x) = \frac{4}{3}\frac{\sqrt{2m}}{e\hbar}\frac{(E_{vac}-E_x)^{3/2}}{\mathcal{E}}$$

from which we have

$$G^{tr}_F(E_F) = \frac{4}{3}\frac{\sqrt{2m}}{e\hbar}\frac{W^{3/2}}{\mathcal{E}} \qquad 6.31a$$

and

$$d_F^{tr} = \frac{\sqrt{8m}}{\hbar} \frac{\sqrt{W}}{e\mathcal{E}}$$

6.31b

Substituting the above expression, 6.31a into 6.25 to obtain $T^{tr}(E_F, 0)$ and then the latter and d_F^{tr}, into the expression for the current, equation 6.30, we get

$$J^{tr} = \frac{e^3 \mathcal{E}^2}{16\pi^2} \frac{\Theta(T)}{\hbar W} exp\left[\frac{-4\sqrt{2m}}{3e\hbar} \frac{W^{3/2}}{\mathcal{E}}\right]$$

6.32

This is the Fowler–Nordheim equation uncorrected for the image potential. It relates the current density to the applied electric field $\mathcal{E}$. Of course, experiments measure the current I, not the current density J. In this simple case of planar geometry, we have

$$I = J^{tr} A$$

6.33

where A is the emitting area. So if a plot of $ln\left(I/\mathcal{E}^2\right)$ v $\frac{1}{\mathcal{E}}$ is made, see figure 6.8, one can

obtain the workfunction from the slope of this curve. This equation was obtained by Fowler and Nordheim in their first paper on Field Emission. As mentioned, experiments at that time indicated the inadequacy of this equation, so Nordheim proceeded with the inclusion of the image potential in the Gamow exponent.

When this is done the resulting integrals are much more complex than the simple case of the triangular barrier, they become linear combinations of the elliptic functions of the first and second kind. However, the mathematics are quite standard. Adopting the superscript

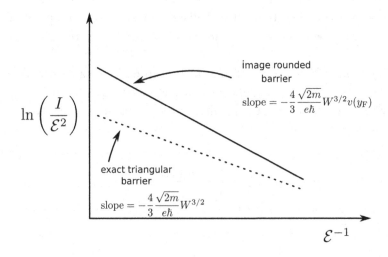

FIGURE 6.8 The FN plot for the exact triangular barrier is linear and for the image rounded barrier almost linear: the latter always gives a higher current.

FN for the image corrected triangular barrier, as is common, after a rather long set of transformations of variables we have

$$G^{FN}(E_x) = G^{tr}(E_x)v(y)$$ 6.34

where

$$y = e\left(\frac{e\mathcal{E}}{4\pi\varepsilon_0}\right)^{1/2}\frac{1}{(E_{vac} - E_x)}$$ 6.35a

and

$$v(y) \approx 1 - y^2 + \frac{y^2 \ln y}{3}$$ 6.35b

Actually $v(y)$ is an elliptic function defined by an integral equation but the above approximation obtained by Forbes [1] is fairly accurate and good enough for our purposes.

From equations 6.34 and 6.35a we deduce that

$$d_F^{FN} = \frac{\sqrt{8m}}{\hbar}\frac{\sqrt{W}}{e\mathcal{E}}\left[v(y) - \frac{2}{3}y\frac{dv(y)}{dy}\right] =$$

$$= \frac{\sqrt{8m}}{\hbar}\frac{\sqrt{W}}{e\mathcal{E}}t(y)$$ 6.36

where $t(y) = v(y) - 2/3\,yv(y)$ with y evaluated at the Fermi level, $y = y_F$ and

$$G^{FN}(E_x = E_F) = G^{tr}(E_F)v(y = y_F)$$ 6.37

Substituting into the basic equation 6.28 we get

$$J = \frac{e^3}{16\pi^2\hbar}\frac{\mathcal{E}^2\Theta(T)}{Wt^2(y_F)}exp\left[-\frac{4\sqrt{2m}}{3e\hbar}\frac{W^{3/2}}{\mathcal{E}}v(y_F)\right]$$ 6.38

It can be observed that equation 6.38 has the same structure as the corresponding equation for the simple triangular barrier with correction factors $v(y_F)$ and $t^2(y_F)$ inserted in the exponent and the preexponential terms respectively. Values for these functions, $v(y)$ and $t(y)$, are available so that again a plot of $\ln(I\mathcal{E}^2)$ vs $1/\mathcal{E}$ - (usually abbreviated "FN plot") can be obtained. Note that y depends on $\mathcal{E}$ so that the FN plot, based on equation 6.38, is not exactly straight but almost straight as an inspection of figure 6.8 shows because both $v(y)$ and $t(y)$ are slowly varying functions of y. Nevertheless the procedure for obtaining the workfuction W, as outlined for the simple triangular barrier, can be carried over to the FN barrier with adequate precision.

6.4 THE 3-DIMENSIONAL WKB PROBLEM

The 3-dimensional WKB problem remains, still, a difficult quantum mechanical problem that has not as yet been solved in its generality. Specific approaches to the problem exist most of which have taken place relatively recently (1980s). The critical reader may ask why we should bother with such a theory when the theory of emission from planar surfaces is adequate. The answer is that emitters nowadays are not planar but are pin-like objects possessing a sharp apex with a radius of curvature R at the top. The radius R was of the order of microns in the 1950s but has continued decreasing ever since and in many applications it is now of the order of nanometers. The prime reason for having pin-pointed emitters is to enhance the applied electric field because sharp metallic features attract electric field lines. Although surfaces of radius of curvature of microns can be considered locally flat, this is no longer possible with an emitter of R in the range of nanometers and a 3-dimensional calculation of the current density is necessary.

Consider a region of space (B), where $E - V < 0$, sandwiched between two regions (A) and (C) where $V - E > 0$, see figure 6.9. As explained in 6.2, any incidence of electrons with an angle away from the normal decreases the transmission coefficient exponentially with that angle and we therefore consider only normal incidence to this barrier, see figure 6.9. In our presentation we follow closely the analysis of Das and Mahanty [2].

Again as in section 6.1, we write for the wavefunction

$$\Psi = exp\left[\frac{i\sigma(r)}{\hbar}\right] \tag{6.39}$$

where Plank's constant $\hbar$ has now been introduced to make the argument of the exponential a pure number. Then, following exactly the steps of section 6.1, we get the 3-dimensional equivalent of equation 6.8

$$(\nabla\sigma)^2 - i\hbar\nabla^2\sigma = 2m\left[E - V(r)\right] \tag{6.40}$$

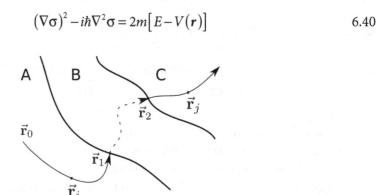

FIGURE 6.9 Region B is a forbidden region sandwiched between allowed regions A and C. A mobile enters region B normal at $\mathbf{r}_1$ and leaves region B at $\mathbf{r}_2$. See text for the calculation of the path.

where E, $V(r)$ are the total energy and potential energy terms respectively of the Schroedinger equation

$$\left[\frac{\hbar^2}{2m} \nabla^2 + E - V(r) \right] \Psi = 0$$

Assuming again that $V(r)$ is slowly varying, in the sense of section 6.1, we can ignore as small the second derivative of σ and obtain an approximate zeroth order solution σ_0

$$(\nabla \sigma_0)^2 = 2m \left[E - V(r) \right] \qquad 6.41$$

The formal solution of 6.41 is a line integral of the form

$$\sigma_0(r) = \sigma_0(r_0) + \int_{l_0}^{l} \sqrt{2m \left[E - V(r) \right]} dl \qquad 6.42$$

The path of integration is the curve of the steepest gradient (descend) of $\sigma_0(r)$. This is also equal to the path of an electron since $\sigma(r)$ is the action of the electron. However, contrary to the 1-dimensional case, the path of the electron or particle in general is not known and this constitutes the major difficulty of the multi-dimensional WKB. Such a path going from the classically accessible region (A) into the classically forbidden region (B) and reemerging from such a region in (C) is shown in figure 6.9. Let r_1 and r_2 be the points of entry to and exit from B respectively. We will come back to the problem of the evaluation of an electron's path, but first let us write, according to 6.42, the wavefunctions of an electron propagating exclusively inside the classically accessible region A and the wavefunction of an electron having gone through the forbidden region B and reemerging in region C. So for an electron propagating from point r_0 to point r_i we have (see figure 6.9) for its wavefunction

$$\Psi(r_i) = exp\left[\frac{i}{\hbar} \left(\sigma_0(r_0) + \int_{l_0}^{l_i} \sqrt{2m(E - V)} dl \right) \right] \qquad 6.43$$

where dl denotes an infinitesimal length along the electron's path. For the wavefunction of an electron at point r_j inside region C, having gone through B, we have

$$\Psi(r_j) = exp\left[\frac{i}{\hbar} \left(\sigma_0(r_0) + \int_{l_0}^{l_1} \sqrt{2m(E - V)} dl + i \int_{l_1}^{l_2} \sqrt{2m(V - E)} dl + \int_{l_2}^{l_j} \sqrt{2m(E - V)} dl \right) \right] \qquad 6.44$$

Note that in the second integral of equation 6.44 we have replaced the negative term $(E - V)$ by the positive $(V - E)$ and have taken outside of the square root the factor (i). Furthermore we have neglected multiple reflections inside region (B) which will exponentially decrease the transmission coefficient.

Our basic equation for the current density J, equation 1.60b of chapter 1, can be written more compactly as

$$J = \frac{\hbar}{m} \text{Im}\left(\Psi^* \nabla \Psi\right) \qquad 6.45$$

Substituting equation 6.43 in 6.45 we get, after some manipulations, that the current density J_A in region (A), is

$$J_A(r) = \sqrt{\frac{2}{m}(E - V(r))} t(r) \qquad 6.46$$

where $t(r)$ is the tangent vector of unit magnitude along the trajectory at r. Similarly using equation 6.44 in equation 6.45 for region (C) we get

$$J_C(r) = \left[\sqrt{\frac{2}{m}[E - V(r)]}\right] \cdot exp(-T_{12}) t(r) \qquad 6.47$$

where

$$T_{12} = \frac{2}{\hbar} \int_{l_1}^{l_2} \sqrt{2m(V(r) - E)} \, dl \qquad 6.48$$

and the line integration from l_1 to l_2 is in region (B). If we now take the ratio of $|J_C(r_2)|$ to $|J_A(r_1)|$ we run into infinities because at the points r_1 and r_2 $E = V$. These infinities can be lifted if we linearize the potential near the turning points (or use L' Hopital's rule) so that we finally get that the transmission coefficient T is of the form

$$T = \left[\frac{(dVdl)_{l=l_2}}{(dVdl)_{l=l_1}}\right] exp(-T_{12}) \qquad 6.49$$

Equation 6.49 does not seem to be a correct generalization of the 1-D case. The only difference from the 1-dimensional (1-D) case is the pre-exponential term which is unity in the 1-D case. The pre-exponential term is usually in the range of 1–3 whereas the exponential term (a genuine generalization of the 1-D case) may vary by orders of magnitude. We will therefore accept 6.49 as a generalization of the 1-D case because the numerical difference from including or not including the awkward pre-exponential factor is insignificant. The cause of this discrepancy is the fact that the WKB expressions for the current density cannot be used at the turning points.

6.5 FIELD EMISSION FROM CURVED SURFACES (ELECTRON GUNS)

As already mentioned in the previous section, field emission rarely takes place from planar surfaces but usually from pin-pointed emitters which can enhance the electric field around them by preferentially collecting electric field lines at their apex. The most common used model (but not the most accurate one) is the so called the hemisphere on a post model shown in figure 6.10. The emitter is simulated as having a semispherical tip of radius R and a main cylindrical body of height h. The emitter is placed on the lower plate of a parallel plate capacitor on the upper plate of which a voltage $V > 0$ is applied. It is assumed that the distance between the plates is much larger than h. This model does not accurately simulate the shape of an emitter but retains all the characteristics necessary to obtain a reliable estimate of the enhancement factor γ, defined by the ratio of the local field $\mathcal{E}_l$ at the apex of the emitter to the macroscopically applied field $\mathcal{E}_M = V/d$ where d is the distance between the plates.

Numerical calculations of the Poisson equation by the present author [3] and Egdcombe and Valdre [4] for the enhancement of the field γ show that

$$\gamma \equiv \frac{\mathcal{E}_l}{\mathcal{E}_M} \approx 0.7 \cdot \frac{h}{R}\left(\text{if } h \gg R\right) \qquad 6.50$$

where $\mathcal{E}_l$ above stands for the local electric field at the apex of the emitter, so the important quantities, as far as enhancement of the field is concerned, are the height h and the curvature at the apex. Obviously different structures will have different enhancement factors, so equation 6.50 above shall only be used as a guide for the order of magnitude to be expected from a given emitting structure. For example, a typical emitter may have a radius of curvature R at the apex of $10nm$ and a typical height of 1 micron so that we expect to obtain experimentally $\gamma \approx 70$. Indeed γ is most of the times treated as a parameter in the theory and is determined experimentally by the Fowler Nordheim plots, provided that the workfunction is known.

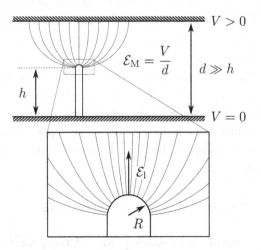

FIGURE 6.10   A pointed emitter of height h inside a parallel plate capacitor of distance d between the plates attracts the electric field lines at its apex, thus enhancing the electric field.

The Fowler–Nordheim (FN) theory for planar surfaces can readily be taken over to the theory of field emission from spherical surfaces if the radius of curvature R at the apex is large enough. What does "large enough" mean? The length of the tunneling region x_{tun} varies from $1-2nm$, so if $R \gg x_{tun}$ the electric field lines near the apex, where most of the current emanates, can be considered parallel to each other, i.e. quasi-1-dimensional. In that case the potential barrier preserves its form discussed in the previous section. Then we can use for the emitted current density at the apex $J(\theta = 0)$- where θ = angle to the emitter axis the FN equation with just $\mathcal{E}_l$ substituted for $\mathcal{E}$ in 6.38, namely

$$J(\theta = 0) = \frac{e^3}{16\pi^2 \hbar} \frac{\gamma^2 \mathcal{E}_M^2}{W t^2 (y_F)} exp\left[-\frac{4\sqrt{2m}}{3e\hbar} \frac{W^{3/2}}{\gamma \mathcal{E}_M} v(y_F) \right] \qquad 6.51$$

As one moves away from the apex, i.e. as θ increases, the normal to the surface electric field diminishes. A very simple way of looking at this is to observe that the highest concentration of electric field lines occurs at the apex. So we can write

$$J(\theta) = J(0)\Theta(\theta) \qquad 6.52$$

where $\Theta(\theta)$ is a decreasing function of θ.

The total current will be given by

$$I = J(0) A_{eff} \qquad 6.53$$

where, contrary to equation 6.33 and because of the spherical geometry here, A_{eff} is an effective area which includes also corrections due to the fact that J is not constant over the surface of the emitter. If the radius of curvature of the emitter R is not much greater than x_{tun} (practically if $R < 20$ nanometers) the above simple approximation of using the FN equation with $\mathcal{E}_l$ as the real field no longer holds. Complications then arise in both the form of the tunnelling potential and in the evaluation of the paths of the electrons that are no longer straight lines. We consider the changes in the form of the tunnelling potential fist. The result we are going to give holds for all kind of emitting surfaces and is not limited to the hemisphere on a past problem. Figure 6.11 portrays an arbitrary shape tip with the inscribed sphere at its apex.

The surface around the apex may be approximated by the surface of the inscribed sphere. Assume spherical coordinates (r, φ, θ) with their centre at the centre of the inscribed sphere. The system has rotational symmetry so φ does not play a role. The potential energy term that will enter the calculation of the Gamow coefficient (equation 6.25) can always be written in the three-dimensional case in the form

$$V_{tun}(r, \theta) = E_F + W - eu_L(r, \theta) - eu_{im}(r) \qquad 6.54$$

where $u_L(r, \theta)$ is the solution of the Laplace equation (having the dimension of volts)

$$\nabla^2 u_L = 0 \qquad 6.55$$

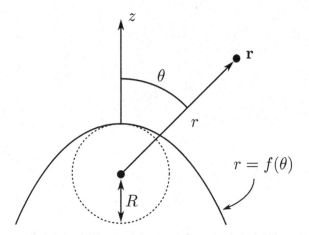

FIGURE 6.11   Geometry for the calculation of the emitted current at the apex of a nanoemitter.

and

$$u_{im}(r) = \frac{eR}{8\pi\varepsilon_0(r^2 - R^2)} \qquad 6.56$$

is the image potential of a sphere. Decomposition 6.54 is valid for any surface. The minus sign in front of the u_L term comes from the charge of the electron. In the case of a planar surface $u_L = \mathcal{E}x$ and $u_{im}(x) = e^2/16x\pi$.

A general but simple expression for $u_L(r, \theta = 0)$ near the surface of the emitter to second order in r can be obtained as follows. The Laplace equation in spherical coordinates is

$$\frac{\partial^2 u_L}{\partial r^2} + \frac{2}{r}\frac{\partial u_L}{\partial r} + \frac{\cot(\theta)}{r^2}\frac{\partial u_L}{\partial \theta} + \frac{1}{r^2}\frac{\partial^2 u_L}{\partial \theta^2} = 0 \qquad 6.57$$

The surface of the emitter however is an equipotential surface, hence all tangential fields are zero on it. Therefore all derivatives with respect to θ vanish. We then get

$$\left.\frac{\partial^2 u_L}{\partial r^2}\right|_{(R,0)} = -\frac{2}{R}\left.\frac{\partial u_L}{\partial r}\right|_{(R,0)} \qquad 6.58$$

If we now make a Taylor expansion of u_L up to second order with respect to $z = r - R$ at $\theta = 0$ we get

$$u_L(z) = u_L(0) + \left.\frac{\partial u_L}{\partial r}\right|_{(R,0)} z + \frac{1}{2}\left.\frac{\partial^2 u_L}{\partial r^2}\right|_{(R,0)} z^2 =$$

$$= u_L(0) + \mathcal{E}z - \frac{\mathcal{E}z^2}{R} \qquad 6.59$$

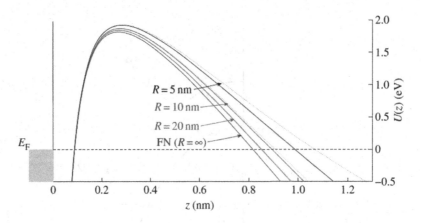

FIGURE 6.12   Potential energy distribution outside a nanoemitter for several values of the radius of curvature R of the nanoemitter. The field at the apex of the tip was kept constant at 5V/nm.

A more rigorous proof of 6.59 is given in Kyritsakis and Xanthakis [5] for any curved surface. The electrostatic potential $u_L(z)$ leads into a tunneling potential energy $V_{tun}(z)$ that an electron sees which is of the form

$$V_{tun}(z)= E_F + W - e\mathcal{E}z - \frac{e^2}{16\pi\varepsilon_0 z\left(1+\dfrac{z}{2R}\right)} + \frac{e\mathcal{E}z^2}{R} \qquad 6.60$$

The variation of this potential energy with respect to R at a constant electric field at the emitter apex is shown in figure 6.12. It can be seen that as R decreases the extent of the barrier increases, thus making tunneling more difficult. Furthermore, it was found that the accuracy of the above expansion becomes inadequate as R approaches the value $R = 5nm$.

The current density along the axis of the tip $J(\theta = 0)$, can now be calculated provided that the evaluation of the awkward integral entering the Gamow exponent can be performed. This is not possible in the general case but a very good approximation can be obtained if the last term of the expansion 6.60 is treated as a perturbation using the Leibnitz integral. This is beyond the scope of this book. The interested reader is referred to reference [5]. It must be made clear, however, that the whole procedure is purely mathematical and no further physical principles are involved. Here we only quote the rather long but in closed form result

$$J(0)= \frac{e^3}{16\pi^2\hbar} \frac{\mathcal{E}^2}{W} \left(t(y_F) + \frac{W}{e\mathcal{E}R} \psi(y_F) \right)^{-2} exp\left[\frac{-4\sqrt{2m}}{3eh} \frac{W^{3/2}}{\mathcal{E}} \left(v(y_F) + \frac{W}{e\mathcal{E}R} w(y_F) \right) \right] \qquad 6.61$$

where $w(y)$ and $\psi(y)$ are new functions defined in terms of $v(y)$ and y is the usual variable of FN theory. It can be seen that equation 6.61 is of a modified FN type: the exponential and pre-exponential FN terms are present in 6.61 but the curvature of the emitter has introduced correction factors in both the pre-exponential and exponential terms given

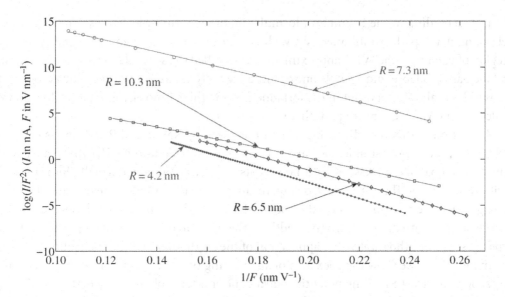

FIGURE 6.13 Fowler–Nordheim plots of a nanoemitter for several values of the radius of curvature R: solid lines denote theoretical values and isolated points experimental data.

respectively by $\psi(y)$ and $w(y)$. The current can now be calculated by a slight generalization and improvement of equation 6.53

$$I = \sigma J(0) A_{eff} \qquad 6.62$$

where σ is a numerical factor that corrects the supply function to take into account that the latter is made out of Bloch functions and not made out of simple plane waves.

The FN diagrams corresponding to equation 6.62 are shown in figure 6.13. The plots are not straight lines but have a slight curvature which is mainly due to the last term (the second order term) in 6.60. From this curvature of the FN plot, the radii of curvature R of the emitters can be obtained by reverse engineering and are shown in figure 6.13. Note that on the logarithmic scale the term σA_{eff} will appear as the intercept on the ordinate axis. This is unknown from theory—for each particular plot in figure 6.13, it has been obtained from experiment. Continuous lines in this figure denote theoretical results, isolated points experimental values, see [5] for more details.

The above theory of curved surfaces (not necessarily spherical) is good enough if one is interested in the total amount of current emitted. If one is interested in the distribution of current in space and the trajectories of the electrons in space or the evaluation of A_{eff} one has to abandon the 1-dimensional WKB and use the 3-dimensional one. One has then to revert to numerical calculations and abandon any hope of an algebraic, analytic manipulation. The main difficulty lies in the evaluation of the paths of the electrons and the corresponding evaluation of the line integrals defining the transmission coefficient, see equations 6.48, 6.49. Of course, no problems occur if these line integrals are evaluated numerically. We must emphasize that paths are not defined in Quantum Mechanics.

In fact, according to the Feynmann formulation of Quantum Mechanics, a particle travels along many paths simultaneously, with each path being assigned a probability. In the region of validity of the WKB approximation, also called the semi-classical approximation, all the paths collapse into a single one or more precisely all are very close to each other and to the classical trajectory. But in the forbidden region there is no classical path. How do we calculate the WKB quantum path then?

A solution to this very difficult problem was given by Kapur and Peierls [6] as early as 1937. The procedure is simple: the barrier $V(r) - E$ is inverted into $E - V(r)$ to create a well instead of a barrier; a classical path now exists if an electron falls normally into the well with zero velocity. This is the electron's quantum path in the forbidden region and the path along which the transmission coefficient is calculated. The 3-dimensional WKB could be very useful in situations where the width of the emanated beam is of importance as in spectroscopy or lithography. The application of the method in practice is as follows.

In figure 6.14 we show the apex area of an emitting tip of arbitrary shape (not necessarily spherical). The tunnelling potential outside the emitter is obtained by first obtaining u_L by solving the Laplace equation either numerically or algebraically and then adding the image potential. Then the two equipotential surfaces where $V = E$ can be calculated and these define the tunnelling or forbidden region. It is shown as the line-shaded area in figure 6.14. The emitter is the inner all shaded semi-circle. For any entry point r_1 on the inner boundary of the forbidden region, the point of exit r_2 on the outer boundary of the tunnelling region is obtained by calculating the entire path in the tunnelling region as described above. The transmission coefficient from r_1 to r_2 is then calculated according to equation 6.49 and then the normally emitted current density $J(r_2)$ is obtained by the standard methods of chapter 5, i.e. by multiplying the supply function by the transmission coefficient and integrating over the normal energy. The assumption behind this choice is a) that the 2- or 3-dimensional problem can be reduced to a set of 1-dimensional problems and b) that locally at each surface element dS a plane wave suppy function is incident on dS. The procedure can be repeated for every point, or more precisely for every surface element dS, on the

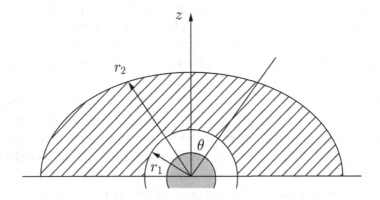

FIGURE 6.14 The tunnelling region (line-shaded area) around a nanoemitter. If the curvature of the emitter at its apex is high, an electron entering the tunnelling region at $\mathbf{r}_1$ will come out at $\mathbf{r}_2$, where vectors $\mathbf{r}_1$ and $\mathbf{r}_2$ do not necessarily lie on the same line.

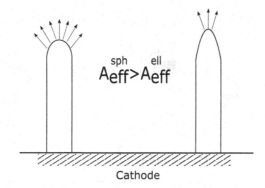

FIGURE 6.15   Two emitters with different R. The emitter with the smaller R emits electrons enclosed within a smaller angle of emission.

inner boundary of the forbidden region, and then $J(r_2)$ can be integrated (over the outer boundary) to obtain the total current I instead of using an A_{eff} obtained from experiment.

Furthermore, the paths in the classically allowed regions can be equally calculated and all of the paths of the electrons can thus be obtained. An important characteristic which arises from such calculations is that the emitting area depends on the apex shape of the emitter. Figure 6.15 shows schematically the beam emanated from an emitter which is nearly spherical at its apex and the beam emanated from a surface which is sharper (say ellipsoidal) at its apex. It can clearly be seen that the beam which emanates from the ellipsoidal surface has less angular spread, a characteristic very useful in practice. This is a general characteristic of surfaces: the sharper is a surface, the less the angular spread of the emitted electrons.

6.6 THE VACUUM TRANSISTOR

The monumental technological progress that has been achieved in the second half of the 20th century and the first two decades of the 21st century can be attributed to a large extent to the micro- and nano-electronics technologies and these, in turn, to the invention of the transistor and the subsequent invention of the integrated circuit whose inventors won Nobel prizes for their work. The transistor was a transformation in solid state form of the vacuum triode or vacuum valve, although the term "transformation" does not exactly do justice to the transistor inventors. It was the possibility of the miniaturization and durability of the transistor that led to the microelectronics and nanoelectronics era and the subsequent technological development. The vacuum valve was pushed during this time into a very limited number of applications. Any transistor (there are a few types) works on the following principle: an electrode injects electrons into a channel made of a semiconductor material which travel along the channel and reach another electrode at the other end of the channel. A third electrode called a gate may or may not impose a barrier to the electron flow depending on the voltage attached to it. Depending on this voltage, electrons may flow or not flow in the channel. The current involved is a drift–diffusion current over the barrier, the sort of current that we discussed in chapter 3.

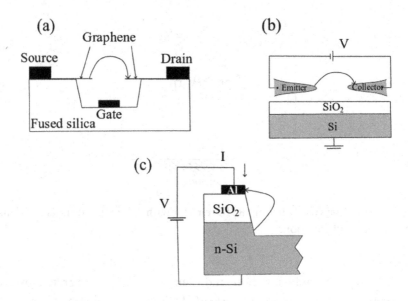

FIGURE 6.16 Various forms of the newly invented vacuum transistor. The flow of electrons in space is shown by an arrow.

Very recently, a so called vacuum transistor has been manufactured by a few laboratories which have replaced the channel with vacuum. Furthermore, the injection into vacuum is accomplished by field emission. So this transistor is a true revival of the vacuum valve with the major difference that it is miniaturized, the cathode–anode distance is several nanometers, and modern lithography is used for its production. Several versions of the vacuum transistor made of semiconductor materials are shown in figure 6.16. The advantage of electron flow in vacuum should be obvious: there are no collisions with a lattice. Furthermore, the device is immune to radiation.

A simplification or idealization of the vacuum transistor for which calculations have been performed by the author is shown in figure 6.17, where two pin-like metallic structures

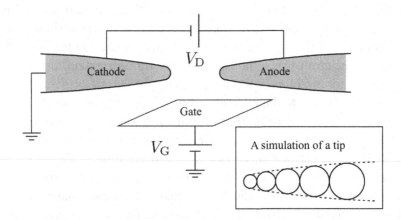

FIGURE 6.17 Picture of a typical vacuum transistor. For the purpose of efficient calculations, the emitting tip is simulated by a stack of spheres as shown in the inset.

act as anode and cathode while a planar metal gate lies above them. The structure as a whole seems to be standing in vacuum while the anode and cathode in real devices lie on a Si/SiO_2 substrate, as shown in figure 6.16b. For the purpose of simplifying calculations we will ignore the difference in the dielectric constant of the substrate and vacuum $(\varepsilon_r(SiO_2) = 3.9)$. This allows us to illuminate the functioning of the vacuum transistor with a relatively simple theory that is accurate enough.

The current I_D between the pin-like structures, the cathode and the anode, is determined by the tunnelling-forbidden region, adjacent to the cathode. This region is created by the applied voltage V_D between anode and cathode but also by the voltage V_G applied to the gate. Under the simplification discussed above, the electrostatic potential around the cathode can be evaluated if we approximate the anode and cathode as piles of spheres as shown in the inset of figure 6.17. Since the potential of an isolated sphere u_{i,sph^-}, where i denotes the site on which the sphere is located, is a well-known problem, the total potential in the space between anode and cathode is just the sum over i of these $u_{i,sph}$ plus the linear one created by the planar gate. More details can be found in [7].

If the electrostatic potential, a solution of the Laplace equation, is evaluated by this indirect method, the tunnelling potential energy $U_{tun}(r,\theta,\varphi)$ can be calculated according to the prescription

$$U_{tun}(r,\theta,\varphi) = E_F + W - eu_L(r,\theta,\varphi) - eu_{im} \qquad 6.63$$

where u_L is the solution of the Laplace equation and u_{im} is the image potential usually taken to be the spherical image potential. Note that we have changed the symbol for the potential energy from V to U since we will need V to denote voltages on electrodes in accordance with standard practice in transistor nomenclature. Due to presence of the gate plane, the system no longer has rotational symmetry, so in spherical coordinates the tunnelling potential energy is a function of φ also. The calculated tunnelling potential energy barrier along the line connecting the apex of the cathode to the apex of the anode (the $\theta = 0$, $\varphi = 0$ direction) for various values of the applied voltage V_G to the gate is shown in figure 6.18. It can be seen that the gate voltage V_G diminishes significantly both the height and length of the barrier, thus allowing a lot more current to flow between anode and cathode.

This is shown in figure 6.19 where we plot I_D with respect to V_D for various values of V_G. The current I_D can be obtained by a surface integration of the current density $J(\theta,\varphi)$ at temperature $T = 0$. The latter has been calculated numerically by obtaining the transmission coefficient along radial lines as outlined in the previous section. It can be seen that V_G has a catalytic effect on the current I_D. The latter increases by orders of magnitude with a small increase by 1–2 volts in V_G. Note that due to the logarithmic scale on the current axis in figure 6.19, the (0,0) point cannot be portrayed. It is rather early (2020) to decide whether the recently developed vacuum transistor will play any role in the field of nanoelectronics. We have presented it here mainly because it illustrates in more concrete term the "transistor action", i.e. the control of the flow of current in a channel by a voltage applied to a third electrode. The whole area of present-day nanoelectronics mainly relies on the operation of

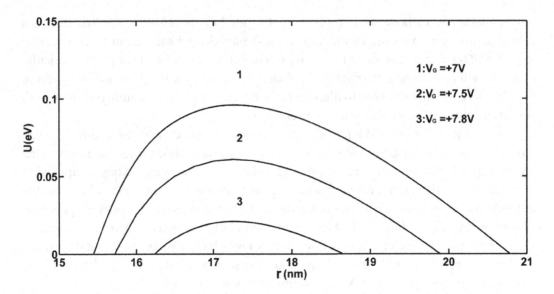

FIGURE 6.18   Potential energy outside the emitter of a vacuum transistor. The gate voltage decreases both the length and height of the barrier.

the MOSFET (the Metal Oxide Semiconductor Field Effect Transistor), a Si-based device. Also, by an increasing degree, possible successors to the MOSFET based on III–V or other materials play a significant role. The operation of the semiconductor FETs is not as simple as the vacuum transistor, so we allocate a separate chapter for the Si devices and another for what is usually termed "post Si" devices or technologies. These form the content of the next two chapters.

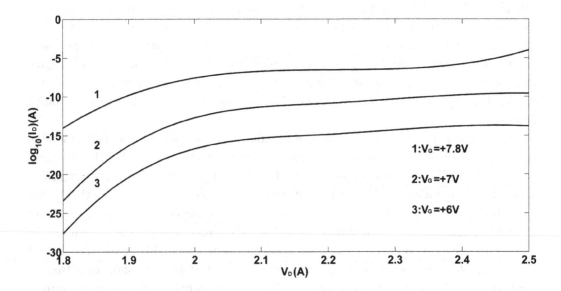

FIGURE 6.19   Current–voltage I_D –V_D characteristic (logarithmic scale) for various values of the gate voltage. The latter seems to be very efficient.

PROBLEMS

6.1 Obtain an equation for σ_2, i.e. the third order exponent, in the WKB expansion of a wavefunction.

6.2 Use the polynomial expansion 6.35b for the (correcting) field emission function $\upsilon(y)$ to estimate the percentage average deviation from the straight line of the Fowler–Nardheim plot of the triangular barrier.

6.3 Derive equations 6.46, 6.47 for the current densities J_A and J_c in the allowed regions in the 3-dimensional WKB problem.

6.4 Show that the enhancement factor of a protrusion in the form of a hemisphere on a cathode is only 3.

6.5 Prove that the enhancement factor γ of a sphere inside a parallel plate capacitor connected with a very thin wire to the cathode (i.e. equipotential to the cathode) is

$$\gamma = c + h/R$$

where h is the height of the centre of the sphere above the cathode, R is the radius of the sphere and c is a constant of the order of 1.

REFERENCES

1. R.G. Forbes Appl. Phy. Lett. 89, 113122(2006)
2. B. Das and J. Mahanty Phys.Rev. B36, 898(1987)
3. G.C. Kokkorakis, A. Modinos, and J.P. Xanthakis J. Appl. Phys. 91, 4580(2002)
4. C.J. Edgcombe and U. Valdre Phil. Mag. B82, 987(2002)
5. A. Kyritsakis and J.P. Xanthakis Proc. R. Soc. A 471, 20140911(2015)
6. R.L. Kapur and R. Peierls Proc. R. Soc. Lond. A 163, 606(1937)
7. M.S. Tsagarakis and J.P. Xanthakis AIP Advances 9, 105314(2019)

CHAPTER 7

The MOSFET

7.1 INTRODUCTION

The number of internet users in 2020 is approaching the 5 billion mark. Each of these users possesses a personal computer (PC) with which he or she communicates with the rest of the world. At the same time, big computers in organizations around the world help solve complex problems like predicting the local weather. In universities, the computer is an everyday tool for both teaching and research. Furthermore, mobile or cellular communication would not exist had it not been for the availability of computing power inside the phone itself and at the installations of every major phone company around the world. At the heart of these technologies (and many more) lies the metal–oxide–semiconductor field effect transistor, MOSFET for short—an electronic switch.

7.2 PRINCIPLE OF OPERATION OF THE MOSFET

A diagram of a MOSFET is shown in figure 7.1a. A P-type (or N-type) Si substrate) has three metal contacts on top of it. The ones to the left and right are called source and drain respectively and are labelled S and D. Below each of the two there is a heavily doped N region (usually symbolized N^{++} or n^{++}) which forms a PN junction with the Si substrate. The third contact, the gate, abbreviated as G in Figure 7.1a, is not in contact with the Si substrate but a thin layer of insulating oxide has been grown between the gate electrode and the Si substrate. This oxide was SiO_2 from the invention of the MOSFET until only a few years ago. It has now been replaced by HfO_2(hafnium dioxide) for the very short-channel MOSFETs now in operation. For the sake of simplicity, we will continue to use the term oxide to include both SiO_2 and HfO_2.

Assume for a moment that no voltage is applied to the gate while a positive voltage V_{DS} is applied between source and drain. The potential energy diagram, i.e. the change of the conduction band minimum as a function of the distance x along the channel, is shown in figure 7.1b. The steps in the potential energy under the source and drain areas are due to the built-in barriers of the PN junction in the respective areas. Electrons from the source electrode cannot flow to the drain electrode because of the barrier that has been formed, so the current I_{DS} is therefore nearly zero. This barrier can be reduced and current I_{DS} may

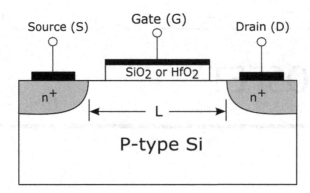

FIGURE 7.1a   Layout of a MOSFET.

flow if a positive voltage, V_{GS}, with respect to the substrate is applied to the gate. If substantial current is to flow this voltage must exceed a certain threshold voltage V_{th}, to be calculated later, i.e. we must have $V_{GS} > V_{th}$. We have already seen such a situation in connection with the vacuum transistor, but what happens in MOSFETs is much more complicated and demands greater analysis.

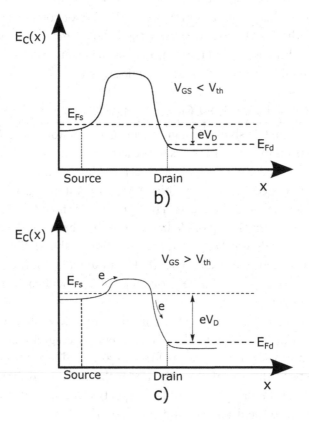

FIGURE 7.1b,c   Variation of the conduction band minimum E_c along the distance from source to drain given a gate voltage V_G which is (a) less than the threshold voltage and (b) larger than the threshold voltage.

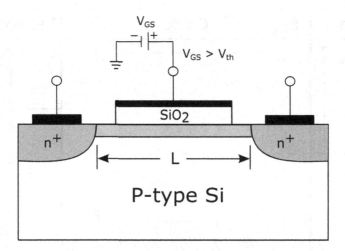

FIGURE 7.1d   Formation of the channel connecting the source to the drain.

When $V_{GS} < V_{th}$, this voltage will push away from the oxide–Si interface the holes, i.e. the majority carriers of the P semiconductor. As the holes move away from the interface, they expose the negative charge of the acceptors and hence create a negative depletion region in the Si semiconductor immediately below the oxide, see figure 7.2a. Upon further increase of the gate voltage such that $V_{GS} > V_{th}$ not only the holes have been pushed away from the oxide/Si interface, but also the electrons, the minority carriers, have been attracted to the interface and have concentrated there, see figure 7.2b. The net effect is that a thin layer in the Si semiconductor, adjacent to the oxide, has changed its character from P-type to N-type and thus a channel of N-type connects the source to the drain and current I_{DS} can flow. This amounts to lowering the barrier as shown in figure 7.1c. The channel is shown schematically in figure 7.1d. We note that in most applications the substrate is short-circuited with the source, so the subscript S in V_{GS} may refer also to the source. Henceforth, therefore we may occasionally omit the second subscript S in V_{GS}, V_{DS} or I_{DS}. The value of the gate voltage V_{GS} required to achieve inversion of the semiconductor under the gate is what we have called V_{th}.

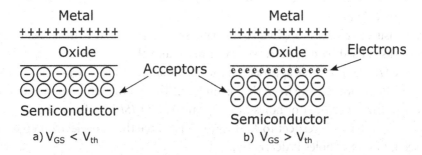

FIGURE 7.2   The two types of changes in the semi-conductor below the gate when the MOSFET is (a) below threshold and (b) above threshold.

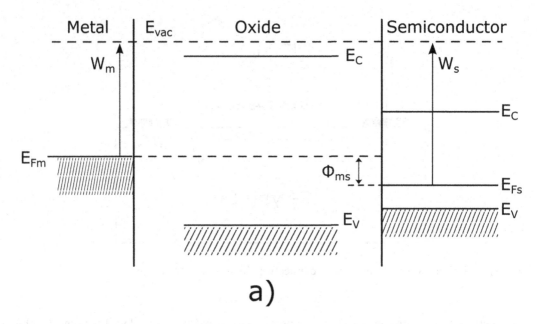

FIGURE 7.3a The band structure of a metal, an oxide, and a semiconductor when they are isolated from each other.

We must now look in more detail at the band diagrams of this structure. Figure 7.3a shows the band diagrams of the components of a MOSFET when isolated from each other. E_{Fm} and W_m stand for the Fermi level and workfunction in the metal, while E_{Fs} and W_s are the corresponding quantities in the semiconductor. The remaining symbols have their conventional meaning. When the three material components are connected to form a MOSFET (we emphasize that this is only, as before, a thought experiment), the Fermi levels of all the component materials will equalize just as in the PN and Schottky junctions that we discussed in chapter 3. The resulting band structure is shown in figure 7.3b. The width of the oxide is denoted by d. For the Fermi levels in the metal and semiconductor to equalize, an exchange of charge must take place between them, so a surface layer of positive charge appears on the metal and a negative layer on the semiconductor. Just as in the Schottky diode, this exchange of charge will create a band bending in the semiconductor which will be equal to $\varphi_{ms} = E_{Fm} - E_{Fs} = W_m - W_s$, as shown in figure 7.3b.

When a positive voltage is applied to the gate, see figure 7.3c, the initial bending downwards that was caused by Φ_{ms} increases even more and the bottom of the conduction E_C of P–Si at the SiO_2/Si interface approaches the common Fermi level whereas the top of the valence E_V of P–Si moves away from the Fermi level. This picture corresponds to the physical process of the movement of holes away from the SiO_2/Si interface. The movement of both E_C and E_V of Si with respect to E_{Fs} locally, guarantee that the electron concentration is increased and that of holes is decreased.

As the voltage V_{GS} is further increased there comes a point where the bottom of the conduction band E_C of Si comes very close to the Fermi level, in which case electrons fill

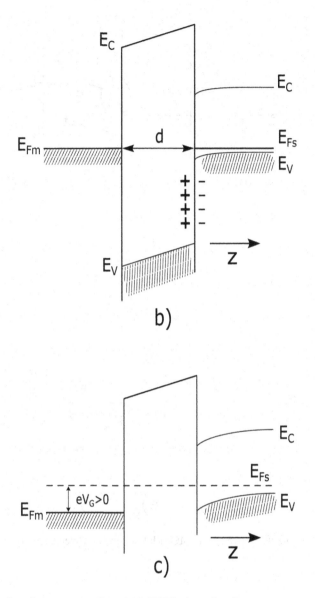

FIGURE 7.3b,c The band structure of a p-MOSFET when the three constituents of figure 7.3a are in contact: (b) no gate voltage V_G, (c) a small $V_G > 0$.

the conduction band of Si substantially while holes are depleted even more, see figure 7.3d, thereby establishing locally the condition $n > p$ which amounts to what we have called inversion. The variation of the various charges as V_{GS} is changed, is shown in figure 7.4. In figure 7.4a, we show the charges present in a MOSFET when $0 \leq V_{GS} < V_{th}$, where V_{th} is the critical value of V_{GS} when inversion has just occurred. A more precise definition and calculation of V_{th} will be given immediately below. Q_m stands for the positive charge per unit area of the surface layer of the gate metal at the metal/SiO_2 interface, Q_n stands for the charge per unit area of the electrons in the conduction band of Si and Q_A stands for the charge per unit area of the depletion region which is due to the exposed (not compensated by holes)

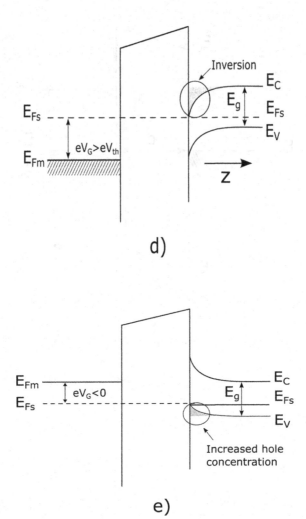

d)

e)

FIGURE 7.3d,e   The band structure of a p-MOSFET when the three constituents of figure 7.3a are in contact: (d) $V_G > V_{th}$ and (e) $V_G < 0$.

charge of the acceptor ions. If l is the width of the depletion region and assuming a fully depleted region we have $Q_A = -eN_Al$.

The situation when $V_{GS} > V_{th}$ is shown in figure 7.4b. It can be seen that Q_n has increased abruptly and significantly due to the exponential nature of the Fermi–Dirac probability distribution, whereas Q_A has increased only marginally. It must be noted that beyond the onset of inversion the width of the depletion does not increase anymore because the essentially planar charge Q_n screens the applied electric field and prevents it from penetrating any further into the Si semiconductor space charge layer. Its width l then attains a maximum value l_{max}. What happens if $V_{GS} < 0$? That situation is simpler. The bands bend upwards, E_V of Si comes closer to the Fermi level, and the concentration of holes near the interface of SiO_2 / Si increases as a result of the physical process of the attraction of holes by the now-negative V_{GS}. The band picture is shown in figure 7.3e and the charges

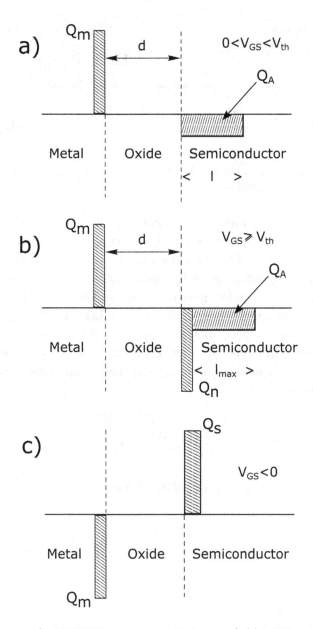

FIGURE 7.4   Charges in the MOSFET structure as V_G is varied: (a) $0 \leq V_{GS} < V_{th}$, (b) $V_{GS} > V_{th}$ and (c) $V_{GS} < 0$.

in figure 7.4c. Q_S in figure 7.4c stands for the surface charge of holes on the semiconductor side of the interface.

A more quantitative definition of V_{th}, the value of the gate voltage at which inversion has just occurred, can be given as follows. In figure 7.5a, we show the position of the Fermi level of a P-type semiconductor. Generally speaking, for a P-type semiconductor to transform into a N-type semiconductor with the same number of electrons as the number of holes it had originally, the Fermi level E_{Fp} must move up the band gap by $2 \cdot (E_{Fi} - E_{Fp})$, with E_{Fi}

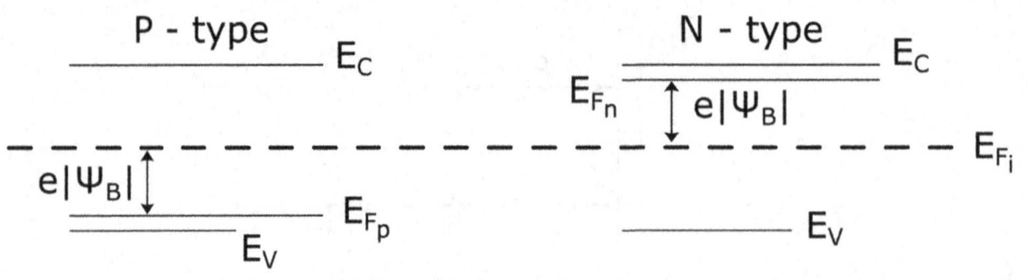

P - type _____ E_C

N - type _____ E_C

E_{Fn} $\uparrow e|\Psi_B|$

- E_{Fi}

$e|\Psi_B| \updownarrow$ _____ E_{Fp}

E_V

_____ E_V

FIGURE 7.5a The energy interval by which the Fermi level of a P-semiconductor has to be lifted to transform it to an N-semiconductor.

being the Fermi level of an intrinsic semiconductor (located approximately at mid-gap). The transformation may be accomplished by, say, extra doping. In the case of a MOSFET, the inversion from P-type to N-type is not accomplished by doping but by band-bending, so the bottom of the conduction band E_C must bend by $V_s = 2 \cdot (E_{Fi} - E_{Fp}) = 2 \cdot \Psi_B$. This is the required voltage drop in Si for inversion. This band bending takes place mainly near the Si / SiO_2 interface and quickly goes to zero in the bulk of Si. This is shown in figure 7.5b. Essentially this has already been shown in figures 7.3d but without any further definitions. Assuming complete depletion of carriers in the band bending region, i.e. assuming $\rho = -eN_A^-$, we get from the one-dimensional Poisson equation in the manner we have done for the Schottky diode

$$V(z) = V_S \left[1 - \frac{z}{l} \right]^2 \qquad 7.1a$$

where z is the direction along the depth of the MOSFET and

$$V_S = V(0) = \frac{eN_A l^2}{2\epsilon_{Si}} \qquad 7.1b$$

with ϵ_{Si} the dielectric constant of Si.

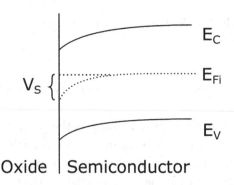

FIGURE 7.5b The variation of the potential along the depth of the Si layer in a MOSFET.

But the gate voltage V_{GS} is applied on the metal electrode. Hence a certain proportion of it is droped in the oxide. Taking into account that there is already a band bending present (equal to Φ_{ms}) at $V_{GS} = 0$ we have for V_{th}

$$V_{th} = \Delta V_{ox} + 2\Psi_B + \Phi_{ms} \qquad 7.2$$

The difference in workfunctions Φ_{ms} between the metal and the semiconductor has an algebraic sign which is negative if the conduction band bends downwards and positive if it bends upwards. From equation 2.63 we have (substituting $p = N_A$ for a P-type semiconductor)

$$\Psi_B = KTln\left[\frac{N_A}{n_i}\right] \qquad 7.3$$

On the other hand, treating the charges on either side of the oxide as a capacitor we have

$$\Delta V_{ox} = \frac{Q_A}{C_{ox}} = \frac{eN_A l_{max}}{C_{ox}} \qquad 7.4$$

where $C_{ox} = \epsilon_{ox}/d$, d is the thickness of the oxide and ϵ_{ox} is the dielectric constant of the oxide.

In equation 7.4 we have assumed that just prior to inversion Q_n can be neglected. Note also that Q_A, Q_n denote charges per unit area and consequently C_{ox} denotes the oxide capacitance per unit area. Note also that the width of the depletion region in Si has attained its maximum value. What is this maximum value? The value that corresponds to a band-bending of $2\Psi_B$. Using the theory of PN junction (see equation 3.45) we have that

$$l_{max} = \sqrt{\frac{2\epsilon_{Si}(2\Psi_B)}{eN_A}} \qquad 7.5$$

so that equation 7.2 becomes

$$V_{th} = 2\Psi_B + \frac{\sqrt{4\epsilon_{Si}eN_A\Psi_B}}{C_{ox}} + \Phi_{ms} \qquad 7.6$$

Equation 7.4 should be viewed as an approximation for the voltage drop in SiO_2 because it has been assumed that the oxide is free of trapped charges inside it and of charged interfacial defects in the SiO_2 / Si interface. However, equation 7.2 remains exact if the voltage differences produced by such charges are included in ΔV_{ox}. It is worthwhile pointing out that if a capacitance measurement is made as V_{GS} is increased, a negative step will be observed

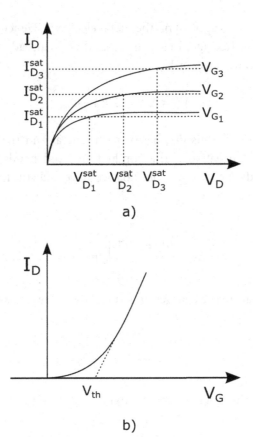

FIGURE 7.6 (a) Typical I_D–V_D characteristics for constant V_G and (b) I_D–V_G characteristic for constant V_D.

at $V_{GS} = V_{th}$. This happens because the capacitance of the MOS structure C before threshold is C_{ox}, whereas after threshold it is given by

$$\frac{1}{C} = \frac{1}{C_{ox}} + \frac{1}{C_d}$$

where C_d is the capacitance of the depletion region, see figure 7.4a–c to see why this is so. The step becomes positive if the substrate is N type.

We now examine qualitatively the dependence of the current I_{DS} on the voltage V_{DS} between source and drain. Assuming a gate voltage such that $V_{GS} > V_{th}$, i.e. that the N-type channel has been created, an increase in V_{DS} initially produces a corresponding linear increase in I_{DS}, see figure 7.6a where the complete characteristics are given. This is easily understood; the electrons flow with a higher drift velocity in the channel. The device acts as a resistor. As V_{DS} is further increased, we observe that the rate of increase slows down and the current eventually almost saturates at $V_{DS} = V_{DS}^{SAT}$.

What physically happens is that as V_{DS} increases, the PN junction below the drain becomes more negatively biased and hence its depletion region widens. It initially decreases

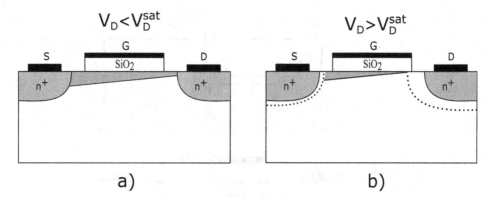

a) b)

FIGURE 7.7   Cross-section of the MOSFET channel (a) below and (b) above saturation voltage. The dotted lines denote the extent of the depletion regions.

the depth of the channel, see figure 7.7a, slowing down the rate of increase of the current, and finally "pinching" the channel. This happens just below the drain. Further increase of V_{DS} beyond V_{DS}^{SAT} moves the "pinching" point towards the source and effectively decreases the length of the channel, see figure 7.7b. A further, very small increase in the current is then observed. The variation of I_{DS} with V_{GS} is given in figure 7.6b. This is easily understood. Below threshold there is no current while above it a higher gate voltage attracts more electrons in the channel and the current increases accordingly. The reader should have revolted by now. If part of the channel is deprived of its carriers (since the channel has been pinched), how come the current remains constant and not drop? We will delve into this matter later but the short answer is that a channel of a very small cross-section survives that can carry the current under the action of very high electric fields inside the "pinched" region of the channel.

7.3 SIMPLE CLASSICAL THEORY

a. **Linear Region, $I_{DS} - V_{DS}$ characteristics**
   A cross-section of a MOSFET in the linear region is shown in figure 7.8. The shortening of the cross-section of the channel at the drain end is intentionally ignored. $V(x)$ is the horizontal potential along the channel from the source to a point A at a distance x from the source. The distance x is measured along the channel, the length of which is L. Obviously, $V(L) \equiv V_{DS}$. The mobile charges, i.e. the electrons in the conduction band of Si, are the ones we have called Q_n in figure 7.4b. From the discussion in the previous section, we know that Q_n is substantial only for $V_{GS} \geq V_{th}$, i.e. $Q_n \sim 0$ below the threshold voltage. Below V_{th}, the applied voltage V_{GS} mainly increases the width of the space charge layer in Si.
   When a voltage V_{DS} is applied to the device, as in figure 7.8, then it is safe according to the discussion above to assume that the mobile charges at the point A are proportional to $\left(V_{GS} - V(x) - V_{th}\right)$ and we can write for Q_n

$$Q_n = C_{ox}\left[\left(V_{GS} - V(x) - V_{th}\right)\right] \qquad\qquad 7.7$$

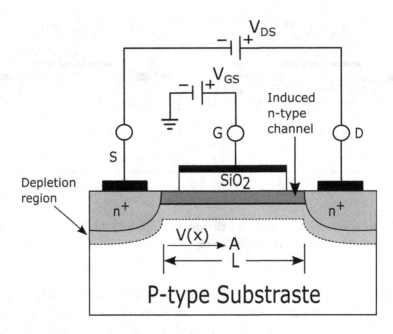

FIGURE 7.8   A MOSFET in operation in the linear region: geometry for the calculation of the current.

The current I_{DS} is then simply

$$I_{DS} = \frac{dQ}{dt} \qquad 7.8$$

where dQ is the differential charge in a differential volume element of length dx. Remembering that Q_n is charge per unit area we have

$$dQ = Q_n W dx \qquad 7.9$$

where W is the width of the device, i.e. the dimension perpendicular to the plane of figure 7.8. The differential time for the charge dQ to travel by the distance dx is

$$dt = \frac{dx}{v_{dr}(x)} \qquad 7.10$$

where $v_{dr}(x)$ is the electron drift velocity at x. If the electric field ε is low, the drift velocity can be put into the form $v_{dr} = \mu_n \varepsilon$ where μ_n is the mobility of the electrons. The latter is not equal to the bulk mobility, but it is approximately half the value of bulk mobility, because of interfacial scattering in the Si/SiO_2 interface. Taking into account that

$$\varepsilon(x) = -\frac{dV}{dx} \qquad 7.11$$

and that the charge Q_n is negative, from equations 7.8–7.11 we get

$$I_{DS} = WC_{ox}\mu_n(V_{GS} - V(x) - V_{th})\frac{dV}{dx}$$

7.12

Separating variables and integrating from the start to the end of the channel we have

$$\int_0^L I_{DS}dx = \int_0^{V_{DS}} WC_{ox}\mu_n(V_{GS} - V - V_{th})dV \Rightarrow$$

$$\Rightarrow I_{DS}^{Lin} = \frac{WC_{ox}\mu_n}{2L}\left[2(V_{GS} - V_{th})V_{DS} - V_{DS}^2\right]$$

7.13

Note that I_{DS} has been taken outside the integral because it is constant along the channel. For values of $V_{DS} \ll V_{GS} - V_{th}$, the term V_{DS}^2 can be neglected and this condition defines a more stringent condition for the extent of the linear region. We note that certain authors define a third intermediate region between the linear and the saturation one in which the quadratic term above is important.

b. **Saturation Region $I_{DS} - V_{DS}$ characteristic**
   As noted previously, when pinch-off occurs the transistor enters the saturation region where the current increases at a very slow rate. Before examining the physical reasons for the near constancy of the current, we will obtain first an expression for the current similar to equation 7.13. We only need to find the value of V_{DS}^{SAT} which separates the linear region from the saturation region. From equations 7.7 to 7.9, it is evident that pinch-off occurs the pinching at point (P) when

$$V_{GS} - V(x = P) - V_{th} = 0$$

7.14

But $V(x = P) = V_{DS}^{SAT}$ by definition (irrespective of whether the point (P) is below the drain or has moved towards the source). Hence

$$V_{DS}^{SAT} = V_{GS} - V_{th}$$

7.15

Since V_{DS}^{SAT} denotes both the end of the linear region and the start of the saturation region, we can substitute in 7.13 for $V_{DS} = V_{DS}^{SAT}$ to get for the current in the saturation regime

$$I_{DS}^{SAT} = \frac{WC_{ox}\mu_n}{2L}(V_{GS} - V_{th})^2$$

7.16

Above this value of I_{DS}^{SAT} the $I_{DS} - V_{DS}$, characteristics show a small linear increase, see figure 7.6. This is due to the decrease of the length of the channel as V_{DS} is increasing above V_{DS}^{SAT} as shown in figure 7.7. Let this decrease be $\Delta L = x_{ch}$. Then the effective length of the channel $L_{eff} = L - x_{ch}$. The current will be

$$I_{DS}^{SAT} = \frac{k}{2} \frac{W}{L_{eff}} (V_{GS} - V_{th})^2 \qquad 7.17$$

where $k = \mu_n C_{ox}$. Since the small increase in I_{DS} beyond V_{DS}^{SAT} is almost linear in V_{DS} we only need to define a constant for dI_{DS}/dV_{DS}. If we write

$$\frac{dI_{DS}}{dV_{DS}} = \frac{I_D}{V_A} \qquad 7.18$$

as a definition of V_A, then the electrical characteristics for the entire saturation region can be written as

$$I_{DS}^{SAT} = \frac{k}{2} \frac{W}{L} (V_{GS} - V_{th})^2 \left(1 + \frac{V_{DS}}{V_A}\right) \qquad 7.19$$

c. **Physical explanation of the saturation region**

In the previous section, we have already noted the apparent contradiction that the channel is completely pinched-off in the saturation region but the current remains almost constant and in fact even increases slightly. A consequence of the continuity equation is that if there is no charge accumulation, the current in a channel must be independent of the particular cross-section used to measure it. So, there cannot be a finite current in the non-pinched part of the channel and no current in the pinched part where there are no mobile carriers. The answer to this apparent dilemma is of course that the channel is not entirely pinched and a "residual channel", i.e. a non-pinched one, is always present below the drain. The situation is depicted in figure 7.9 where it is shown that the channel consists of two parts; one non-pinched roughly below the source and over the gate and another residual roughly below the drain. The one below the drain has a very small but finite cross-section that can sustain the uniformity of the current.

The possibility of a "pinch-off" of the channel is only a mathematical fiction that essentially derives from a 1-dimensional view of the electrostatics of the problem. In a 1-dimensional view of the Poisson equation, if the depletion layer of the PN junction below the drain becomes equal to the minimum distance between gate and drain the inescapable conclusion is that the N-channel disappears because of "pinch-off". But the problem is not 1-dimensional as in the PN junction because of the crossed fields produced by V_{DS} and V_{GS}. A 2-dimensional solution of the Poisson equation is necessary to give a picture as the one shown in figure 7.9.

Furthermore, taking into account only the 2-dimensional nature of the electrostatics is not sufficient to produce the above-mentioned picture. One must take also

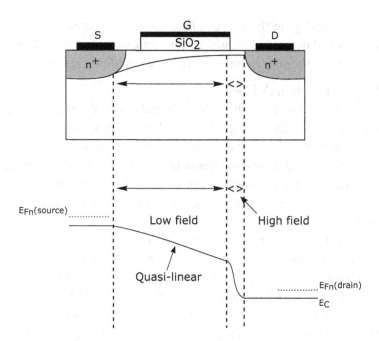

FIGURE 7.9 The channel region is divided into two sub-regions: one in which the electric field is low and the potential almost linear and another in which the field is high and in which "pinching" of the channel has occurred.

into account the saturation of the drift velocity with electric field as discussed in section 4.8. In fact, the start of the "residual channel" appears at the distance from the source where the longitudinal electric field equals the electric field that produces velocity saturation. In other words, all electrons inside the residual channel travel at the v_{dr}^{sat} speed discussed in chapter 4. The potential along the residual channel, however, is not constant, in fact it increases at a much faster rate than the roughly linear one prevailing in the non-pinched part of the channel. This is shown in figure 7.9. An initial quantitative analysis of the formation of the "residual channel" was given by Grebene and Ghandhi in 1969 [1], although its existence was postulated much earlier by W. Shockley, one of the inventors of the transistor. As we will see in the next section, the 2-dimensional Poisson equation with no approximations for the charge density can only be solved in conjunction with the continuity and the Schrödinger equations as a system of equations. To analyze the residual channel, Grebene and Ghandhi had to make many approximations and therefore we will not present their analysis. In the next section, we will show how the 2-dimensional potential, charge density, and current density can be deduced by a system of differential equations. The main results of Grebene and Ghandhi are summarized in figure 7.9.

d. **Ohmic contacts to source and drain**

When metal contacts are made on the source and drain, a metal–semiconductor junction is created, which according to our discussion in section 3.9 is a Schottky diode, i.e. a rectifying junction. No transistor can work having a rectifying diode under the source or drain. Currents need to go in both directions. The problem is resolved by over-doping the semiconductor side of the metal–semiconductor junction, as explained in section 3.9, which substantially increases the tunnelling current from the metal to the semiconductor when the junction is reverse biased.

e. Significance of L, Moore's Law

The value of the channel length L has a particular significance for the device performance; it is the major factor in determining how quickly the device responds, i.e. the speed of the device. In fact, the driving force behind Moore's Law is this connection. More than 50 years ago G. Moore, one of the founders of INTEL, predicted that transistor length will shrink every 18–24 months, and this prediction came true. Apart from the obvious advantage of having more memory in the same chip, this miniaturization led to higher device speed because the maximum frequency of operation of a MOSFET f_T was proportional to $1/L^2$ in the initial stages of miniaturization and $1/L$ later. There is a solid theory of the time dependent MOSFET operation behind all these that we will present later in this chapter, but the basic argument is simple. The response of the device will depend on the time it takes for an electron to transverse it, i.e. travel the distance L of the channel. With devices operating in the saturated velocity regime nowadays, this is equal to L/v^{sat}, and the inverse of that gives the $1/L$ behavior. Devices no longer work at the maximum allowed frequency of a MOSFET due to power dissipation limitations of the whole chip but miniaturization continues according to Moore's Law. The shortening of the channel length L has however introduced new phenomena some of which can be described within the simple classical theory that we have given so far, without resorting to the systems of differential equations that we are going to give in the next section.

f. Short channel and drain induced barrier lowering (DIBL) effects

If one numerically solves the 2-dimensional Poisson equation (under the approximation of the charge density equal to $\rho = -eN_A$) in two cases a) for a short channel (say $L = 50nm$) MOSFET and b) a long channel (say $L = 1\mu m$) MOSFET, one obtains the following result for the variation of the surface potential V_S (cf equation 7.1), shown in figure 7.10, In this figure, V_S refers to the Fermi level of the source and not to the intrinsic Fermi level of the semiconductor. It can be seen that for a long channel MOSFET the surface potential is constant along the channel, as indeed we have assumed implicitly in our analysis so far. Furthermore, the field due to V_{DS} is appreciable only near the drain end of the channel. On the contrary in the short channel MOSFET the surface potential is not constant and changes along the channel, therefore one should denote it by $V_S(x)$. Furthermore it is reduced by a finite amount from the long channel value. What is this difference due to? In the short channel FET, the source and drain extend significantly in length compared to the channel and are in close proximity to each other and so their depletion regions have a significant effect on the channel and hence on the surface potential. This effect leads to the appearance of a maximum in the potential, as shown in figure 7.10.

The net effect is that the threshold voltage V_{th} is now a function of the channel length. This is usually called the short channel effect. Furthermore, V_{th} is reduced by

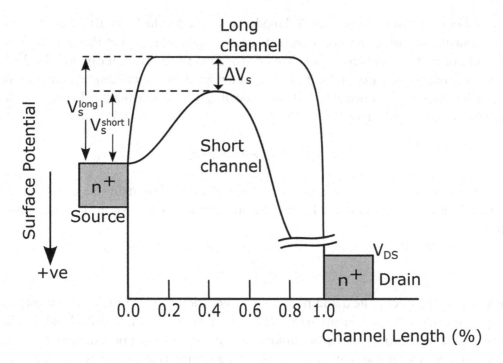

FIGURE 7.10 Variation of the surface potential along the length of its channel for a long and a short channel MOSFET.

the application of V_{DS}. This is called the drain induced barrier lowering (DIBL) effect. If we call V_{th}^o the threshold voltage of a given FET of channel length L at $V_{DS} = 0$ we can write the empirical formula

$$V_{th}(L) = V_{th}^o(L) - AV_{DS} \qquad 7.20$$

where A is independent of V_{DS}. In addition $V_{th}^o(L)$, as the symbol indicates, varies with L when the latter approaches the sub-micron and nano ranges. From the discussion in the previous paragraph it should be clear that the simple approach presented so far is not adequate and we proceed to a more advanced methodology for the analysis of the MOSFET.

7.4 ADVANCED CLASSICAL THEORY

The following section is not a detailed description of the numerical methods of device physics for the MOSFET but it gives the basic equations that are necessary for a complete classical description of devices with no approximations. It also sets the ground for the description of devices in terms of quantum transport. The numerical solution of these equations is a major subject in itself, but it does not fall within the scope of the present book. There are many books on numerical calculations in general.

So far, any analysis we have performed of semiconductor devices has relied on either simplifying the Poisson and continuity equations (as we did for the PN and Schottky diodes) or on devising a semiempirical model (as we did for the MOSFET). Rigorous classical calculations which do not necessitate approximations exist, but they lead to a system of differential equations rather than one equation which by itself can give the device response. In previous uses of the Poisson equation for the PN and Schottky junctions the depletion layer was assumed to be fully depleted; this is an approximation. In principle, the charge density ρ is equal to

$$\rho = e(N_D^+ + p - N_D^- - n) \qquad 7.21$$

Furthermore, since any device, not just the MOSFET, is made of many materials with different dielectric constants ϵ, the Poisson equation should be written in the more general form

$$\nabla \epsilon \nabla \mathcal{E} = -\rho \qquad 7.22$$

where $\mathcal{E}$ is the electric field and ρ is given by equation 7.21. It should be immediately apparent that the Poisson equation cannot be solved by itself because both $\mathcal{E}$ and the electron and hole densities n and p are unknown. Equation 7.22 can of course be combined with equations for n and p to form a system of equations as already stated. Actually, the Poisson equation is a necessary, indispensable equation in all models we are going to present.

The simplest system of differential equations for an N-channel MOSFET is produced when the Poisson equation is complemented by the continuity equation for electrons at steady state (cf. equations 3.33 and 3.34)

$$\nabla J_n = eR_n \qquad 7.23$$

where J_n is given by the drift-diffusion model in 3 dimensions

$$J_n = ne\mu_n \nabla V + eD_n \nabla n \qquad 7.24$$

and R_n is the recombination rate for electrons. Note that in the channel the electrons are the majority carriers and hence the full Schottky-Reed-Hall theory must in principle be used for R_n. However, for unipolar devices as the MOSFET, i.e. for devices operating with mainly one type of carrier, the recombination rate is small and is many times omitted.

From the discussion in section 7.2 it should be obvious that when the substrate is P-type, the channel is N-type and vice versa. Therefore, the above set of equations is not sufficient for a domain of solution where the character of the semiconductor changes from P-type to N-type. The density of holes p must be included in the system of equations and a third equation becomes necessary. We note, however, that if the domain of solution is restricted to the channel, a solution is possible by the above restricted system of 2 equations with 2 unknowns. A flow chart for the solution of this simple 2x2 system of differential equations is shown below.

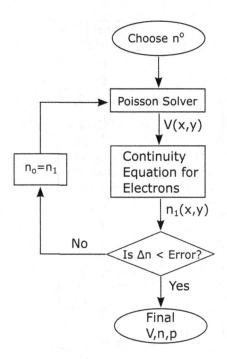

FLOWCHART 1

A much better model is obtained if all V, n, p are included but the electrons are treated with a higher accuracy than the holes. This can be accomplished by treating the electrons by the Boltzmann equation (cf. equations 4.3–4.7) and the holes by the continuity equation. Thus with $f(k,r)$ being the distribution function for electrons, in the notation used in chapter 4, we have for the Boltzmann and continuity equations, 7.25 and 7.26 respectively

$$\frac{1}{h}\nabla_k E(k)\nabla_r f(k,r) + \frac{q}{h}\nabla_r V(r)\nabla_k f(k,r) = \frac{\partial f(k,r)}{\partial t}\Big|_{coll} \qquad 7.25$$

where q= −e and

$$\nabla\big[e\mu_p p(r)\nabla V(r) - eD_p \nabla p(r)\big] = R(n,p) \qquad 7.26$$

To accomplish self-consistency, the distribution function must also appear in the Poisson equation. This is evidently achieved by writing for the electron density $n(r)$ including spin

$$n(r) = \frac{1}{4\pi^3}\int f(k,r)d^3k \qquad 7.27$$

The above system of equations can be solved numerically by discretizing all the unknown functions or by numerically solving the Poisson equation on the one hand and solving the Boltzmann by expanding in the spherical harmonics $Y_{lm}(\theta,\varphi)$ [2]. A flow chart for the solution of the set of equations just described, is shown below.

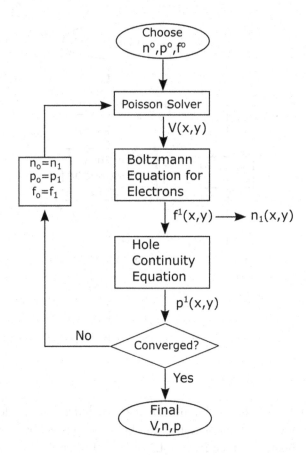

FLOWCHART 2 f^o denotes here an initial guess of the non-equilibrium distribution function and not the value at equilibrium.

The next step in the modelling of the MOSFET is the inclusion of the Schrödinger equation. The inclusion of the latter in the modelling of the MOSFET was found necessary because of the continued miniaturization of the MOSFET and the corresponding effect on the charge density in the transistor channel. In particular, along with the shortening of the channel length, there comes a corresponding reduction of the oxide thickness. This in turn increases the electric field perpendicular to the $Si/oxide$ interface with the result that the bending of the conduction band near the interface becomes very steep, see figure 7.11. Then the electrons in the conduction band of Si cannot be considered to be in a conventional 3-dimensional semiconductor, but in a 2-dimensional one, parallel to the $oxide/Si$ interface with the third dimension, the one perpendicular to the interface, being of nanometric length. The wavefunction of the electrons in the channel experience quantization effects in their density of states, as described in chapter 5. Then a 2-dimensional electron gas (2DEG) is said to be created. The situation is depicted in figure 7.11, where we show the potential well perpendicular to the $Si/oxide$ interface and the corresponding wavefunctions for the first 3 bound levels of the well. Although the above described situation constitutes a

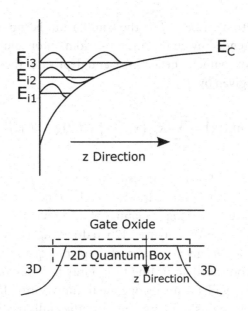

FIGURE 7.11 Quantization of the electrons in the region below the oxide due to the surface potential in the semiconductor. The integer i stands for valley.

quantum effect, we include it here in the classical simulations section, because the transistor equations used in such models for the current density **J** are still classical.

The Schrödinger equation for the direction perpendicular to the oxide surface reads

$$\frac{-\hbar}{2}\frac{d}{dz}\left(\frac{1}{m^*_{i,z}}\frac{dX_{i,j}}{dz}\right)+U_{1D}(z)X_{i,j}=E_{i,j}X_{i,j} \qquad 7.28$$

where $X_{i,j}$ is the (envelope) wavefunction and $E_{i,j}$ are the eigenvalues for the i_{th} valley in the j_{th} sub-band. The term $m^*_{i,z}$ is the effective mass of the i_{th} valley along the z direction which is perpendicular to the $Si/oxide$ interface. Note that we use U for potential energy to avoid confusion with V reserved for potential, i.e. the solution of the Poisson equation. Remember that the relation between the two is U= −eV. The 1-dimensional potential energy $U_{1D}(z)$ can be obtained from the 2-dimensional Poisson equation. The latter can be solved numerically, on a mesh, in a sub-domain called quantum box, as shown in figure 7.11. Equation 7.28 is then solved along any of the lines of the mesh that are perpendicular to the $Si/oxide$ interface by imposing zero boundary conditions on both ends. Along any such line $U_{1D}(z)$ can be extracted from the solution of the Poisson equation.

The electron density $n(z)$ along these perpendicular lines to the interface can be computed from the relation

$$n(z)=\sum_{ij}\left|X_{i,j}(z)\right|^2\int f_0(E)g^{2D}(E)dE \qquad 7.29$$

with g^{2D} given by equation 5.12 and f_0 is the equilibrium Fermi–Dirac distribution. The above model can be extended by augmenting the domain of solution of the Schrödinger equation to 2 dimensions, namely the length x and depth z of the channel. Then the electron density is likewise given by

$$n(x,z) = \sum_{ij} \left| X_{i,j}(x,z) \right|^2 \int f_0(E) g^{1D}(E) dE \qquad 7.30$$

with g^{1D} given by equation 5.19. The peak electron concentration along the channel thus calculated is shown in figure 7.12a, figure redrawn from [3]. One can see that the electron concentration does not behave in the simple manner of the previous section. The current I_{DS} as a function of V_{DS} and V_{GS} is shown in figure 7.12b, redrawn again from [3]. Details about the calculation and a flow chart are given in the next chapter on quantum wells, see flow chart 3.

As miniaturization continued according to Moore's law, the ballistic regime of operation is reached. Present-day FETs already operate in this regime. The Boltzmann equation becomes simplified in the ballistic regime because the collision term is absent from the equation. We will allocate more space to the description and solution of this Boltzmann equation because a) it leads naturally to the quantum description and b) the solution can be inferred from physical intuition and no numerical work is necessary in this case. A lot of the physical intuition actually derives from the Landauer theory of chapter 5. A Landauer

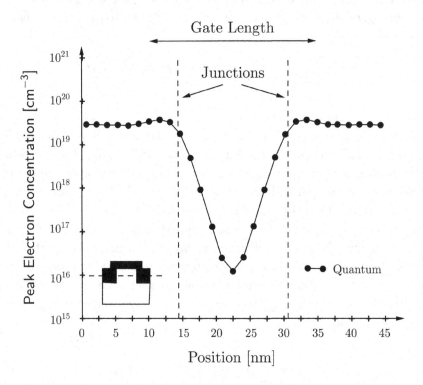

FIGURE 7.12a Calculations of the peak electron concentration in the channel by Pirovano et al [3] using the 2-dimensional Schroedinger equation and a classical transport model.

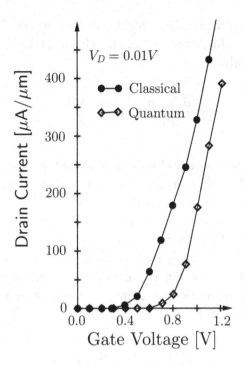

FIGURE 7.12b Calculations by Pirovano et al [3] of the current I_D as a function of gate voltage V_G using the 2-dimensional Schroedinger equation and a classical transport model. For comparison a classical calculation where the Schroedinger equation is not used is also shown.

theory of the MOSFET will be given in the next section. The solution of the Boltzmann equation given here is due to M. Lundstrom [4] and we follow it closely.

The MOSFET channel is considered bounded by two oxide layers, one above it, one below, and connected to two reservoirs, the source and the drain as shown in figure 7.13. Injection of carriers from both the source and drain reservoirs into the channel takes place as the Landauer theory assumes. Let the corresponding Fermi energies be E_{Fl} and E_{Fr}. Obviously

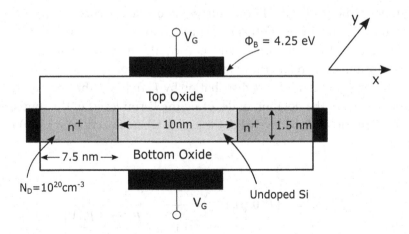

FIGURE 7.13   Layout of the nanodevice used for the Boltzmann calculations by Rhew et al [4].

thermodynamic equilibrium prevails inside the leads-reservoirs. The Boltzmann equation in the steady state of the ballistic regime (i.e. no collisions) takes the form, assuming variation in one dimension only,

$$v_x \frac{\partial f(x, \boldsymbol{p})}{\partial x} - e\mathcal{E}_x \frac{\partial f(x, \boldsymbol{p})}{\partial p_x} = 0 \qquad\qquad 7.31$$

where v_x and $\mathcal{E}_x$ are the velocity and electric fields along the channel, i.e. along the x-direction and f denotes the distribution function in the ballistic regime. Note that the Plank constant in the denominator of the second term has been absorbed in the derivative. For simplicity we assume that all the electrons in the channel occupy only the lowest sub-band of the conduction sub-bands shown in figure 7.11.

Furthermore, the electrons, see figure 7.13, are assumed to be free to move in the (x, y) plane with a single effective mass, that of the transverse effective mass m_t^* of equation 2.46. The momentum vector $\boldsymbol{p}$ spans the (x, y) plane, i.e. $\boldsymbol{p} = (p_x, p_y)$ and the energy $E(x, \boldsymbol{p})$ is in an effective mass approximation

$$E(\mathrm{x}, \boldsymbol{p}) = \frac{p_x^2 + p_y^2}{2m_t^*} + E_c(x) \qquad\qquad 7.32$$

where $E_c(x)$ is the lowest conduction sub-band minimum. This quantity is obtained by solving the Poisson equation in the (x, z) plane self-consistently with the 1-dimensional Schroedinger equation, along z as explained in previous paragraphs. Then the electric field is

$$\mathcal{E}_x = \frac{1}{e} \frac{dE_c(x)}{dx} \qquad\qquad 7.33$$

The boundary conditions for the Boltzmann equation can be specified by assuming that injection into the channel from both source and drain occurs according to Fermi–Dirac statistics. Since both the equilibrium and non-equilibrium distribution functions will appear in our formulae, we retain the notation of chapter 4 on the Boltzmann equation and denote the equilibrium Fermi–Dirac distribution by f_0 and by f the non-equilibrium one. Furthermore we retain the notation of denoting the Fermi levels of the source and drain by E_{Fl} and E_{Fr} (left, right) respectively For the source where injection (at $x = 0$) necessitates $p_x > 0$ we have

$$f(x, \mathbf{p}) = f_0\big(E(0, \mathbf{p}), E_{Fl}\big) = \cfrac{1}{1 + \exp\left(\cfrac{E(0, \mathbf{p} - E_{Fl})}{KT}\right)} \qquad\qquad 7.34$$

and for the drain where injection (at $x = L$) into the channel necessitates $p_x < 0$

$$f(x,\mathbf{p}) = f_0\big(E(L,\mathbf{p}), E_{Fr}\big) = \frac{1}{1 + exp\left(\dfrac{E(L,\mathbf{p} - E_{Fr})}{KT}\right)} \qquad 7.35$$

The occupation of the $(x,\mathbf{p})$ states in the channel can now be inferred by physical intuition, as mentioned earlier. The main idea is that the injected electrons in the channel remain in equilibrium with the reservoir from which they were injected depending on their momentum and position in the channel. To expound this idea, we draw $E_c(x)$, the lowest conduction sub-band profile, together with the respective Fermi levels in figure 7.14. $E_c(x)$ shows a maximum E_{max} at x_{top} near the source. We also show in this figure the parabolic band structure for the electrons at an arbitrary point in the channel. The dependence of $E(x,\mathbf{p})$ on $\mathbf{p}$ may look simple, but it is good enough for our arguments.

Consider the electrons ejected from the source. Those which will climb over the barrier, i.e. those with $p_x > p_{max}$ where

$$p_{xmax} = \sqrt{2m_t^*(E_{max} - E_c(x))} \qquad 7.36$$

are shown with a solid line on the band structure in figure 7.14. All the other electrons injected from the source with $p_x < p_{max}$ will be reflected back. Consider now the electrons injected from the drain. These will have initially a negative p_x. The ones whose energy is below the barrier, i.e. those with $|p_x| < p_{max}$ will travel at most up to x_{top} and will then be reflected back to the drain. These states are marked with a broken curve on the band

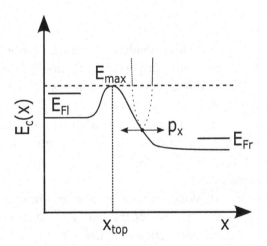

FIGURE 7.14   Variation of the conduction band minimum E_c (x) as a function of the distance along the channel of the nanodevice in figure 7.13. Also shown at an arbitrary point along the channel is the dispersion relation E-k. Electrons from the source which go over the barrier are marked with a solid curve on the dispersion relation, electrons from the drain which also go over the barrier are marked with a dotted curve and those from the drain or source which return back with a broken line.

structure diagram. Finally, the electrons from the drain emitted with negative momentum and energy greater than E_{max} will also climb the barrier and are marked with a dotted curve in figure 7.14. The ballistic solution of the Boltzmann equation can now be written down by putting the electrons in two categories: those above the barrier and those below it.

For $|p_x| > p_{xmax}$

$$f(x,\mathbf{p}) = \begin{cases} \dfrac{1}{1+\exp\left(\dfrac{E(x,\mathbf{p}-E_{Fl})}{KT}\right)}, & \text{if } p_x > p_{xmax} \\[4em] \dfrac{1}{1+exp\left(\dfrac{E(x,\mathbf{p}-E_{Fr})}{KT}\right)}, & \text{if } p_x < -p_{xmax} \end{cases}$$

7.37a

and for $|p_x| < p_{xmax}$

$$f(x,\mathbf{p}) = \begin{cases} \dfrac{1}{1+exp\left(\dfrac{E(x,\mathbf{p}-E_{Fl})}{KT}\right)}, & \text{if } 0 < x < x_{top} \\[4em] \dfrac{1}{1+exp\left(\dfrac{E(x,\mathbf{p}-E_{Fr})}{KT}\right)}, & \text{if } x_{top} < x < L \end{cases}$$

7.37b

From the above discussion it should be evident that the behaviour of $f(x,\mathbf{p})$ at the top of the barrier $x = x_{top}$ is of paramount importance. Figure 7.15 shows cross-sections of $f(v_x, v_y = 0)$ for several drain voltages, where v denotes velocity, at $x = x_{top}$. It shows clearly that as V_{DS} is increased, an asymmetry in the occupation of the velocities in the x-direction appears, hence current begins to flow in the channel. The zero current in the case of $V_{DS} = 0$ is a result of two opposite and equal currents. The voltage V_{DS} spoils this symmetry. Saturation at $V_{DS} = 0.6$ V is also observed. 2-dimensional diagrams of this situation are shown in figure 7.16.

The evolution of $f(v_x, v_y = 0)$ along the channel is shown in figure 7.17 for $V_{DS} = 0.6$ V, i.e. at saturation. At $x = 2.5nm$, i.e. well inside the source, the distribution is very similar to either that of figure 7.15 or to a cross-section of figure 7.16 at $v_y = 0$. As the wavepackets or ballistic electrons travel along the channel, they gain energy from the field and hence their peak velocity moves to higher velocities. Also, their distribution becomes narrower. As we approach the drain, we find the drain-injected electrons which have a symmetric (with respect to positive and negative v_x) distribution since they are reflected back well before reaching x_{top}.

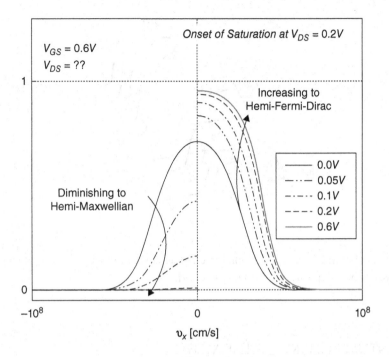

FIGURE 7.15 The 1-dimensional cross-sections of $f(v_x, v_y=0)$ for several V_{DS} at the top of the barrier x_{top} for V_G=0.6V. Note that saturation begins at V_{DS}=0.2V and is complete at V_{DS}=0.6V. Figure has been reproduced from reference [4].

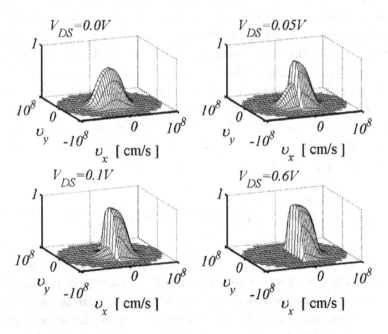

FIGURE 7.16 The 2-dimensional plots of $f(v_x, v_y)$. Note that the distribution of v_y retains the equilibrium shape. Figure has been reproduced from reference [4].

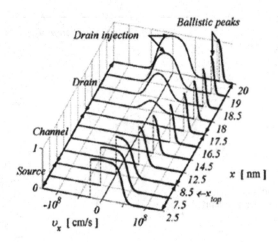

FIGURE 7.17 Evolution of f (v_x,v_y=0) along the channel at saturation (V_{DS}=0.6V). See text for comments. Figure has been reproduced from reference [4].

7.5 QUANTUM THEORY OF THE MOSFET

We now come to a quantum theory of the MOSFET. As already noted, present-day MOSFETs have channel lengths of the order of tens of nanometers or even smaller, so according to the introductory discussion of chapter 5 a classical particle approach is no longer valid. Furthermore, the ballistic transport, expounded in the previous section, must be a necessary ingredient of such a theory. We emphasize again that the use of the Schrödinger equation, although necessary, is not adequate to characterize a methodology as quantum. It is the use of a quantum transport method to analyze the device that does so. As usual, the effective mass equation will be used. We follow closely the analysis of K. Natori [5], one of the pioneers in the subject. This analysis is not only useful for MOSFETs, but can be applied to many devices, so we will use it in the next chapter for devices made out of semiconductors other than Si. We assume here the coordinate system used in the previous section and throughout this chapter. The device is homogenous in the y direction and of width W. The coordinate x is along the channel length and z is along the channel depth.

For a Si channel in a MOSFET the effective mass equation reads

$$\left[\frac{-\hbar^2}{2m_x^*}\frac{\partial^2}{\partial x^2} - \frac{\hbar^2}{2m_y^*}\frac{\partial^2}{\partial y^2} - \frac{\hbar^2}{2m_z^*}\frac{\partial^2}{\partial z^2} + U(x,y,z)\right]\Psi(x,y,z) = E\Psi(x,y,z) \qquad 7.38$$

Since the device is uniform in the y direction, the potential energy along y can be taken as that of a very wide quantum well of width W and of a constant deep potential, as shown in figure 7.18a. The potential energy along x will be of the same form as that used in the solution of the ballistic Boltzmann equation in the previous section. This is shown in figure 7.18b. The maximum at x_{top} of U along x is shown and this specific point will play a major role in the quantum theory of the MOSFET as it did in the solution of the Boltzmann equation.

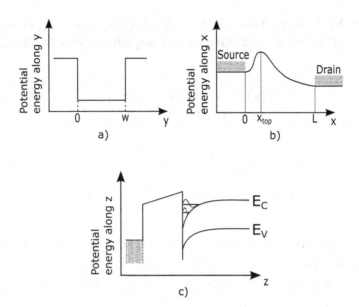

FIGURE 7.18   Potential energy distributions of an electron in a MOSFET along the three spatial directions (a) along its width y, (b) along the channel length x, and (c) along its depth z.

Finally, the potential energy along the depth z of the channel will have the usual triangular-like barrier form shown in 7.11 which is narrow enough as to quantize the eigenfunctions in z direction. We reproduce it here in figure 7.18c for completeness.

We assume that the wavefunction can be written in the following form

$$\Psi(x,y,z) = X(x)\left(\sqrt{\frac{2}{W}}\sin\left(\frac{n_y \pi y}{W}\right)\right)\Phi_{n_z}(x,z) \qquad 7.39$$

where W is the width in the y-direction and n_y, n_z are the quantum numbers in the y and z direction respectively. Note that equation 7.39 cannot be obtained by the method of separation of variables because the triangular barrier shown in figure 7.18c is x-dependent. However since the potential along x is slowly varying compared to z, the $\Phi_{n_z}(x,z)$ can be taken as the solutions of the following quasi-1-dimensional equation

$$\left[-\frac{\hbar^2}{2m_z^*}\frac{\partial^2}{\partial z^2} + U'(x,z)\right]\Phi_{n_z}(x,z) = \epsilon_{n_z}\Phi_{n_z}(x,z) \qquad 7.40$$

where the coordinate x is treated as a parameter. This Schroedinger equation is solved along 1-dimensional cuts perpendicular to the plane of the gate. In 7.40 above the position of the line cut is explicitly shown through the x dependence. Furthermore, the valley index does not appear explicitly. However, the valley index will appear later in the expression for the current to account for the proper counting of the number of electrons emitted from the source and from the drain.

The function $X(x)$ can be obtained from the WKB approximation discussed in chapter 6. From equation 6.17 we get for $X(x)$ for a wave propagating from source to drain

$$X(x) = \frac{C}{\sqrt{p(x)}} \exp\left(\frac{i}{\hbar} \int^x p(x')dx' \right)$$

7.41a

where

$$p(x) = \sqrt{2m_x \left[E - \frac{\hbar^2}{2m_y^*}\left(\frac{n_y \pi}{W} \right)^2 - \epsilon_{n_z}(x) \right]}$$

7.41b

and C is a constant. Note that $n_y\pi/W$ acts like a wavevector in the y direction and the eigenvalues in the y and z direction act as an effective potential in the WKB expression. Around the critical maximum x_{top} equation 7.41 may be approximated by a single plane wave as

$$X(x) = \frac{C'}{\sqrt{\hbar k_{top}}} e^{ik_{top}x}$$

7.42a

where

$$\hbar k_{top} = p(x_{top})$$

7.42b

To calculate the current I we use the Landauer theory of chapter 5 with slight modifications which are necessary here. As usual, we assume that 1) injection from both source and drain takes place so that the net current is the sum of two currents travelling in opposite directions and 2) the electrons injected from the source remain in thermodynamic equilibrium with the reservoir of the source and so do the electrons injected from the drain. On the other hand, we take into account here the different valleys of Si when summing over all the eigenstates normal to the propagation direction and we also take into account that the final state of any electron must be empty. We do not care what happens to the electrons as they cross the channel in each of the two directions. Instead of examining the particular scattering events we simply multiply each beam by the transmission coefficient, which is the same for both directions, to obtain the number of electrons reaching the drain from the source and vice-versa.

The flow of electrons in each direction is given by their velocity times their density times the electron charge (e). The electron density is given by the 1D density of states times the respective (source or drain) Fermi–Dirac distribution. If E_{FS} and E_{FD} denote the Fermi levels in the source and drain respectively and the equilibrium Fermi–Dirac distribution is denoted by $f(E_F, E)$ (as in chapter 5) we have for the current flowing from source to drain

$$I_{S \to D} = e \sum_l \sum_{n_y} \sum_{n_z} \int v_x g_{1D}(E_x) f(E_{FS}, E)(1 - f(E_{FD}, E)T(E_x)dE_x$$

7.43

where the term in parenthesis in 7.43 guarantees that the receiving state is empty and the summation over l indicates summation over valleys. Similarly, for the current injected from the drain and reaching the source we have

$$I_{D \to S} = e \sum_l \sum_{n_y} \sum_{n_z} \int v_x g_{1D}(E_x) f(E_{FD}, E)(1 - f(E_{FS}, E)) T(E_x) dE_x \qquad 7.44$$

Evidently

$$I_{DS} = I_{S \to D} - I_{D \to S} \qquad 7.45$$

The 1-dimensional density of states g_{1D} given by equation 5.20 is not appropriate here because the present potential $U(x,z)$ does not resemble in any way the confining potential of section 5.2b. The function g_{1D} of equations 7.43 and 7.44 can be computed easily: the number of states per spin per dk_x interval is $1/\left(\dfrac{2\pi}{L}\right)$ so that the number of states per dk_x interval per length is $1/\pi$ and from the identity

$$\frac{dN}{dE_x} = \frac{dN}{dk_x} \frac{dk_x}{dE_x}$$

we get

$$g_{1D}^x(E_x) = \frac{1}{\pi} \frac{dk_x}{dE_x} = \left(\pi \hbar \frac{dE_x}{dp_x}\right)^{-1} \qquad 7.46$$

The velocity v_x in the effective mass approximation used here is

$$v_x = \frac{dE_x}{dp_x} = \frac{p_x}{m^*_x}$$

so that the product of the first two factors in 7.43 and 7.44 gives the constant $(\pi \hbar)^{-1}$. Then equation 7.45 reduces to

$$I_{DS} = \sum_l \sum_{n_y} \sum_{n_z} \frac{e}{\pi \hbar} \left[f(E_{FS}, E) - f(E_{FD}, E) \right] T(E_x) dE_x \qquad 7.47$$

We remind the reader that if the Fermi level at the source reservoir is E_{FS}, the Fermi level at the drain reservoir is $E_{FD} = E_{FS} - eV_{DS}$.

In accordance with our results of the Boltzmann equation of the previous section the electron beam that will reach the opposite reservoir must have energy $E_x > E_{n_z}(x_{top})$. We therefore assign a transmission coefficient of $T = 1$ for these energies and a transmission coefficient $T = 0$ for the ones below this limit. Obviously, tunnelling through the barrier is neglected. Actually, tunnelling through the barrier is the ultimate limiting process for the good operation of a MOSFET. The summation over n_y can now be turned into an integration over the normal energy E_y since the width W is usually large. We can therefore write

$$I_{DS} = \frac{e}{\pi\hbar}\sum_l\sum_{n_z}\int g^y{}_{1D}(E_y)dE_y * \int [f(E_{FS},E)-f(E_{FD},E)]dE_x \qquad 7.48$$

The density of states g^y_{1D} will also be given by an equation equivalent to 7.46 (with x substituted by y). This will give a term of the form $E_y^{-1/2}$. Performing the inner integration (which will give $E_y^{1/2}$) we finally get (after changing the variable of integration from E to E/KT (KT is the Boltzmann thermal energy)

$$I_{DS} = \frac{e\sqrt{2m_y}W(KT)^{3/2}}{\pi^2\hbar^2}\sum_l\sum_{n_z}F_{1/2}\left(\frac{E_{FS}-\epsilon_{n_z}(x_{top})}{KT}-F_{1/2}\left(\frac{E_{FD}-\epsilon_{n_z}(x_{top})}{KT}\right)\right) \qquad 7.49a$$

where $F_{1/2}(a)$ is the Fermi-Dirac integral of order $1/2$ defined by

$$F_{1/2}(a) = \int_0^\infty \frac{\sqrt{z}}{1+e^{z-a}}dz \qquad 7.49b$$

and as usual $E_{FD} = E_{FS} - eV_D$. This definition of $F_{1/2}$ is equivalent to the one we gave in chapter 2.

The mobile charge Q_n per unit area at the bottleneck $x = x_{top}$ can be evaluated in the same manner we have evaluated N_L and N_R of equation 5.67. Both contributions from source and drain must be considered. Hence, we can write summing over all sub-bands and valleys

$$|Q_n(x = x_{top})| = \frac{eKT}{2\pi\hbar^2}\sum_l\sum_{n_z}\sqrt{m_x^*m_y^*}\ln\left\{\left[1+\frac{exp\left(E_{FS}-\epsilon_{n_z}\left(x_{top}\right)\right)}{KT}\right]*\left[1+\frac{exp\left(E_{FD}-\epsilon_{n_z}\left(x_{top}\right)\right)}{KT}\right]\right\} \qquad 7.50$$

The current I_{DS} as a function of Q_n is shown in figure 7.19, figure reproduced from Natori (2008) [6]. Two curves are shown, the one labelled as EOSA denotes calculations with an Effective One Sub-Band Approximation, i.e. by considering only the lowest sub-band, the other labelled MSM denotes calculations with many sub-bands. The $I_{DS} - V_{DS}$ characteristics are plotted in figure 7.20 and are compared with the measured characteristics of a 70nm MOSFET. It can be seen that the theory of ballistic transport does not accurately represent the experimental curves. The primary reason for the disagreement is that the approximation of a transmission coefficient equal to one is too extreme for such a gate length of a channel made out of Si. Actually, the result given by equation 7.49 should be considered as the limiting case to which all FETs tend as the channel length L diminishes. We note of

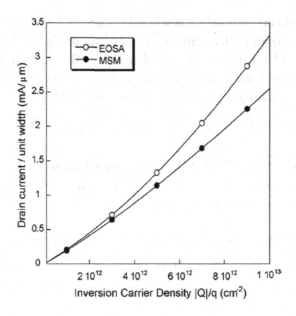

FIGURE 7.19 Current I_{DS} as a function of the mobile charge Q_n at $x=x_{top}$. EOSA stands for Effective One Sub-band Approximation and MSM denotes a calculation including many sub-bands. The figure is reproduced from Natori (2008) [6].

course that in such a limit the current I_{DS} is independent of L. Present-day MOSFETs are nearly ballistic. It is worthwhile noting that III–V FETs, which we are going to examine in the next section, become ballistic at longer channel length than Si MOSFETs because the mean free path λ of the III–V compounds are longer than that of Si, hence the condition $L \ll \lambda$ is realized sooner i.e. at longer L. We now give a simpler theory of semi-ballistic FETs that is probably more physically appealing. The method is due to Lundstrom [6, 7].

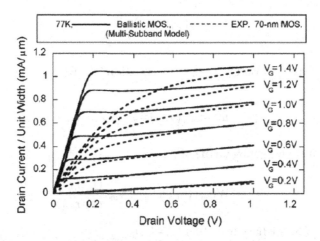

FIGURE 7.20 Comparison of the theoretical (Natori) I_D-V_D curves with experimental curves of a 70 nm MOSFET. Reproduced again from [6].

Unlike the case of long-channel FETs where the current I_{DS} is extracted by an integration over the channel length (c.f. equations 7.7 to 7.13), in the case of nanoFETs the current I_{DS} is due entirely by what happens at the bottleneck x_{top}. The electrons that will manage to overcome the barrier will reach the drain. The point x_{top} is very close to the source so that equation 7.7 at the beginning of the chapter can be rewritten here as

$$Q_n\left(x_{top}\right)=C_{ox}(V_{GS}-V_{th})$$ 7.51

i.e. we have neglected any small effect along the channel due to V_{DS}.

From the discussion so far and the solution of the Boltzmann equation in the previous section we have that in saturation

$$\frac{e-flux\ reaching\ drain}{per\ unit\ width} = \frac{incident\ flux\ to\ x_{top}}{per\ unit\ width}* \frac{percentage\ transmitted\ over\ barrier}{per\ unit\ width}$$

or

$$J_D = J_S t_{ch}$$ 7.52

But the density of electrons $n(z)$ at the bottleneck with z being the depth direction of the channel is equal to the sum of the incident and reflected fluxes divided by the velocity $v(x_{top})$ at that point.

Hence

$$n(z)=\frac{J_s+r_{ch}J_s}{v(x_{top})}=\frac{J_s(1+r_{ch})}{v(x_{top})}$$ 7.53

where r_{ch} stands for the reflection coefficient and then J_D becomes by substituting for J_s from 7.53

$$J_D =n(z)v(x_{top})\frac{t_{ch}}{1+r_{ch}}$$ 7.54

The integral of $n(z)$ is however $Q_n(x_{top})$, so by equation 7.51 we get

$$\int n(z)dz = Q_n\left(x_{top}\right)=C_{ox}(V_{GS}-V_{th})$$ 7.55

and for the current I_{DS}^{sat} we get by virtue of 7.54

$$I_{DS}^{sat} = WC_{ox}v(x_{top})(V_{GS}-V_{th})\frac{(1-r_{ch})}{(1+r_{ch})}$$ 7.56

where in equation 7.56 we have used that $t_{ch}=1-r_{ch}$.

It is now time to draw some conclusions about the differences of long channel FETs and short-nanometric channel FETs. The obvious difference exemplified by the above equation

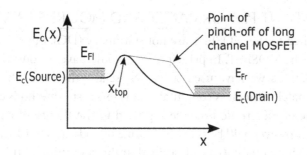

FIGURE 7.21 The critical point in a long and in a short channel MOSFET. In the long channel MOSFET it is the point of pinch-off and is near the drain, while in a short channel MOSFET it is near the source and is the point where the maximum of the barrier is located.

is that the saturation current I_{DS}^{sat} is not proportional to $(V_{GS} - V_{th})^2$, but to $(V_{GS} - V_{th})$. This is a direct consequence, as can be seen by an examination of equations 7.51 to 7.56, of treating current as a wave rather than as a flow of particles. Furthermore, the mechanism of the saturation of the current has changed drastically from a physics point of view. Saturation of the current I_{DS} in the long-channel FET occurs at the pinch-off region which is near the drain side of the channel. The basic mechanism is the saturation of the velocity. On the contrary, saturation of the current in the short-nanometric MOSFET occurs near the source side of the channel and is due to the saturation of the flow over the barrier maximum, i.e. at the point we have called x_{top}. In figure 7.21 we compare the potentials along the channel for short and long L. The point at which the high field region inside the channel begins to be established is indicative of the mechanism of the current saturation. In the long L MOSFET it is near the drain whereas in the short L MOSFET it is near the source.

We now give some approximate expressions for the transmission coefficient that rely more on physical intuition than a strict quantum mechanical theory. These may be used in equation 7.47 if one wishes to include the transmission coefficient in the Natori formula. We differentiate between the linear and saturation regimes because in each region a different length plays the major role. In the linear regime the transmission coefficient $T_L(E)$ can be represented simply by

$$T_L(E) = \frac{l(E)}{L + l(E)} \qquad 7.57$$

where $l(E)$ is the energy dependent mean free path of the channel. Obviously $T_L(E) \to 1$ if $L \ll l(E)$. On the other hand, a different length scale is important in the saturation regime: it is x_{top}, the distance from the source to the bottleneck point along the channel. The reason for this is the following: at saturation there is a large electric field in the channel that is located between x_{top} and the drain, see figure 7.21. If the electrons manage to overcome the barrier at x_{top}, the probability of being returned back to the source by back-scattering is almost zero. Therefore, we can write that the saturation regime transmission coefficient $T_S(E)$ is

$$T_S(E) = \frac{l(E)}{l(E) + x_{top}} \qquad 7.58$$

7.6 TIME-DEPENDENT PERFORMANCE AND MOORE'S LAW

If the voltages applied to the MOSFET are not stationary (DC), but time dependent (AC), so is the current in the MOSFET. In principle, one would have to solve the time dependent version of the equations we have used so far, especially the time dependent continuity equation, but fortunately other less computationally expensive methods exist which belong more to the regime of Electronic Engineering than to the theory of charge transport. A short introduction, however, will be given here. This introduction will serve only to extract Moore's law that we have qualitatively described in the beginning of this chapter.

Consider the simplified circuit diagram shown in figure 7.22. At the gate a DC and AC voltage generator are applied. This will create DC and AC components in all major quantities associated with the MOSFET. We adopt the following convention.

DC part of quantities: capital letter, capital subscript

AC part of quantities: small letter, small subscript

Total quantities: capital letter, small subscript

We then write for DC and AC currents and voltages

$$I_{ds}(t) = I_{DS} + i_{ds}(t) \qquad\qquad 7.59a$$

$$V_{gs}(t) = V_{GS} + v_{gs}(t) \qquad\qquad 7.59b$$

$$V_{ds}(t) = V_{DS} + v_{ds}(t) \qquad\qquad 7.59c$$

If the magnitude of the AC components is small (which normally is the case), a linear Taylor expansion of the above time dependent drain current can be made. Therefore

$$I_{ds}(t) = I_{DS} + \left.\frac{\partial I_{ds}(t)}{\partial V_{gs}}\right|_{V_{GS}} v_{gs}(t) + \left.\frac{\partial I_{ds}(t)}{\partial V_{ds}}\right|_{V_{DS}} v_{ds}(t) \qquad\qquad 7.60$$

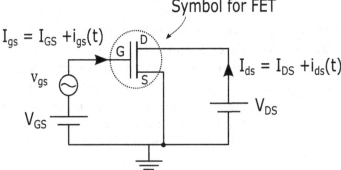

FIGURE 7.22 A simplified circuit diagram for the derivation of the small-signal equivalent circuits.

If the variation in $v_{gs}(t)$ and $v_{ds}(t)$ is slow, that is if the frequencies involved are low, then the derivatives in 7.60 can be equated to their DC values, that is

$$\frac{\partial I_{ds}(t)}{\partial V_{gs}} \simeq \frac{\partial I_{DS}}{\partial V_{GS}} \quad and \quad \frac{\partial I_{ds}(t)}{\partial V_{ds}} \simeq \frac{\partial I_{DS}}{\partial V_{DS}} \qquad 7.61$$

It is important to note that equations 7.61 hold only for low frequencies. When DC measurements of an I–V characteristic are made, it is always assumed that the system has enough time to relax between one measurement and the next so the DC derivatives of I_D can not in principle be used when the system is continuously varying. In essence "low" frequencies means that the system (MOSFET) can follow the input signal with no time delay. At higher frequencies capacitive effects do delay the MOSFET and these effects have to be taken into account.

The methodology of equation 7.60 and 7.61 proceeds therefore in two stages: first the approximation of equation 7.61 is used and a simple circuit, mathematically equivalent to equations 7.60 and 7.61, is derived for low frequencies; then additional capacitors are added to it to account for the actual delays in the MOSFET. All the required information is contained in the AC parts of the quantities of interest, so we only need a circuit that relates these components only. This is called "small signal equivalent circuit" and is just a topological representation of the relations between the AC components of 7.59. The small signal, low frequency circuit equivalent to equations 7.60 and 7.61 is shown in figure 7.23a. Note that the derivative

$$\frac{\partial I_{ds}(t)}{\partial V_{gs}} = \frac{\partial i_{ds}}{\partial v_{gs}} \simeq \frac{\partial I_{DS}}{\partial V_{GS}} = g_m = transconductance$$

is represented as a current generator whereas the derivate

$$\frac{\partial I_{ds}(t)}{\partial V_{ds}} = \frac{\partial i_{ds}}{\partial v_{gs}} \simeq \frac{\partial I_{DS}}{\partial V_{DS}} = \frac{1}{r_o} = output\ conductance$$

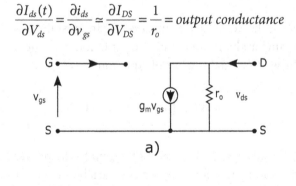

a)

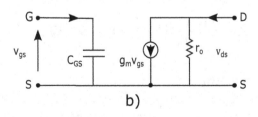

b)

FIGURE 7.23   Small-signal equivalent circuits for (a) low frequencies and (b) high frequencies.

as a resistor. It can easily be seen that the gate is open-circuited because there is no gate current. However, as the input frequencies are increased, AC electric currents can cross dielectrics and so for higher frequencies a capacitor labelled C_{gs} is added between the source and the gate electrodes. So, the small signal equivalent circuit at high frequencies is given in figure 7.23b. Additional capacitances are needed to have a proper representation of the MOSFET time dependent behaviour, but we will stop at this stage because figure 7.23b is adequate in order to extract Moore's Law. But first, we have to find expressions for the elements appearing in figure 7.23b.

From equation 7.19 we have

$$g_m = \frac{\partial I_{DS}}{\partial V_{GS}} = k\frac{W}{L}(V_{GS} - V_{th})\left(1 + \frac{V_{DS}}{V_A}\right) \approx k\frac{W}{L}(V_{GS} - V_{th}) \qquad 7.62$$

We also have from 7.19

$$\frac{1}{r_o} = \frac{\partial I_{DS}}{\partial V_{DS}} \approx \frac{I_{DS}}{V_A} \Rightarrow r_o \approx \frac{V_A}{I_{DS}} \qquad 7.63$$

Note that the above parameters g_m and r_o depend on the DC values of the current and gate voltage.

Next, for the high frequency small signal circuit we have to calculate C_{gs}, i.e. the capacitance between the gate and the source. In the linear region this is easy because we can imagine that the charge of the channel is shared equally between the following two pairs of electrodes: a) gate with source and b) gate with drain. Hence, we can write

$$C_{gs}^{Lin} = \frac{C_{ox}WL}{2} \qquad 7.64$$

In the saturation region, however, where most of the interest lies, the drain has very little contact with the channel and a more detailed calculation is needed. The total charge Q_{tot} below the gate can be written, making use of equation 7.7, as

$$Q_{tot} = WC_{ox}\int (V_{GS} - V(x) - V_{th})dx \qquad 7.65$$

where, we remind the reader, $V(x)$ is the part of V_{DS} up to the point x, measured from the source. Using also equation 7.12 and transforming variables from y to V we get

$$Q_{tot} = \frac{W^2 C_{ox}\mu_n}{I_D}\int_0^{V_{GS}-V_{th}} (V_{GS} - V - V_{th})^2 \, dV \qquad 7.66$$

Performing the integration, we get

$$Q_{tot} = \frac{2}{3}WLC_{ox}(V_{GS} - V_{th}) \qquad 7.67$$

so that the required capacitance is

$$C_{gs} \equiv \frac{\partial Q_{tot}}{\partial V_{GS}} = \frac{2}{3} WLC_{ox}$$ 7.68

We note that in both equations 7.65 and 7.66 the upper limit of integration is the point of pinch-off. Substituting the above values of g_m, r_o and C_{gs} in the small signal equivalent circuit of figure 7.23 gives a very approximate picture of the performance of a MOSFET. More capacitances are needed for an accurate description. However, for the purposes of obtaining Moore's Law it is adequate.

From figure 7.23b we have

$$i_{in} = j\omega C_{gs} v_{gs}$$ 7.69

Furthermore, omitting r_o which is small

$$i_{out} = g_m v_{gs}$$ 7.70

Therefore

$$\left| \frac{i_{out}}{i_{in}} \right| = \left| \frac{g_m}{j\omega C_{gs}} \right|$$ 7.71

A MOSFET is useful for digital circuits if given an input signal (i.e. i_{in}) of digital "1" we get a response (i.e. i_{out}) of also "1", that is expression 7.71 should not become less than one. Given that 7.71 is a decreasing function of ω, this condition sets the maximum allowed frequency for the MOSFET usually denoted by f_T. Substituting the expressions for g_m and C_{gs} we have derived, we get

$$2\pi f_T = \frac{3}{2} \frac{\mu_n}{L^2} (V_{GS} - V_{th})$$ 7.72

This led to the famous Moore's Law. Gordon Moore predicted, based on 7.72, that the number of MOSFETs in a chip would double every one or two years and the frequency of operation would increase accordingly. The law proved correct for several decades, but the corresponding increase in frequency came to a stagnation because of heat dissipations: as the frequency is increased, heat is dissipated faster and there is a danger of chip malfunction or burning. Furthermore, as the length of the channel is decreased to nanometric lengths f_T scales as $1/L$ and not as $1/L^2$. The primary reason for that is that the drift velocity saturates and is no longer proportional to the electric field, see more in the next chapter.

As the length of the channel was decreased the oxide thickness was also decreased to preserve the gate control on the channel. At one point in time the oxide thickness was between 1nm and 2nm. Appreciable tunnelling currents were developing that ruined the performance of the MOSFET. The obvious solution was the use of oxides of higher dielectric constant. Hence the traditional SiO_2 was replaced by HfO_2 which has a relative dielectric constant of approximately 25.

7.7 THE FINFET, A 3-DIMENSIONAL MOSFET

We have seen in this chapter that Si MOSFETs suffer from short channel effects, the lowering of the V_{th} with channel length L for a given L, and the drain induced barrier lowering effect, the dependence of V_{th} on drain voltage. Both effects degrade the performance of the transistor. Common to both cases is the loss of control of the channel by the gate. With the planar structure that we have considered so far there is not much that one can do given the replacement of SiO_2 by HfO_2. A way forward from this problem is to increase the number of gates that surround a channel as shown in figure 7.24. By etching Si, a Si fin is produced which is then surrounded by a dielectric first and then by a metal, thus producing a tri-gate FET. Actually, the FinFET as it is called is more of a double-gate FET than a tri-gate FET for the following reason: the top gate does not have a significant area when compared to the other two gates, and hence this gate is less effective.

The equations governing transport in the FinFET within the classical description (which includes the Schroedinger equation) are no different than the ones we have already presented [7]. They have to be solved of course in three dimensions and that can only be performed numerically and with greater difficulty than for the conventional MOSFETs. However, within the approach of a FinFET as a double-gate FET there are what are called compact models, models that perform some initial approximations so that a fast solution is

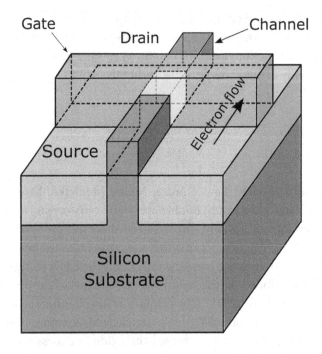

FIGURE 7.24   A Fin-FET in three-dimensions. The gate surrounds the channel from three sides.

obtained. Such compact models may assume that the acceptor doping is low [8] or assume other simplifications. The reader should have little difficulty in following these compact models given the information presented so far in this book. The increased performance of FinFETs is shown in the following figure 7.25, which portrays the difference in leakage currents (figure 7.25a) and gate delays (figure 7.25b) between the conventional planar geometry and the fin geometry of Si MOSFETs.

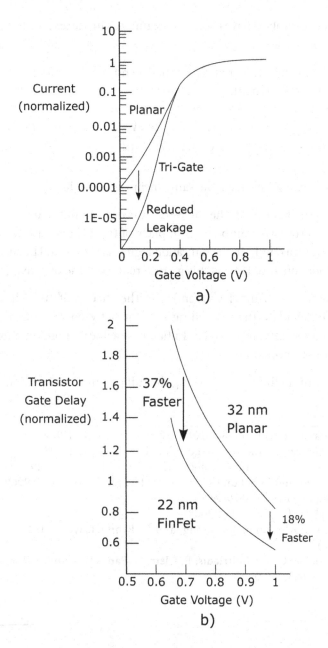

FIGURE 7.25 Superior performance of the Fin-FET compared to the planar one: both (a) the sub-threshold current and (b) the gate delay are smaller for the Fin-FET.

PROBLEMS

7.1 Show that the capacitance of a MOS structure is

$$C_{min} = \cfrac{\varepsilon_{ox}}{d + \cfrac{\varepsilon_{ox} l_{max}}{\varepsilon_s}}$$

where the above symbols have the same meaning as in the text. Justify any intermediate formula you may use.

7.2 Prove the following relation at $V_{DS}=0$ between the gate voltage V_{GS} and the capacitance C of the MOS structure

$$\frac{C_{ox}^2}{C^2} = 1 + \frac{2\varepsilon_{ox} V_{GS}}{e\varepsilon_s N_A d^2}$$

where the above symbols have the same meaning as in the text.

7.3 The transconductance g_m of the MOSFET was derived in the text on the assumption that the DC derivatives can be used in place of the AC derivatives. Assume that the time dependent current I_{ds} in the saturation regime can be found by substituting $V_{GS}+u_{gs}$ for V_{GS} in equation 7.16. How is this g_m different from the one given in the text?

7.4 (Contact Resistance-Tunnel Current) Use the theory of tunnelling developed in chapter 6 to calculate the tunnelling current between the metal contact and the heavily doped source in reverse bias. Initially use a linear approximation for the barrier. Ignore the image potential.

7.5 Calculate the tunnelling current using the full potential of equation 3.75.

REFERENCES

1. A.B. Grebene and S.K. Ghandi. Solid State Electronics 12, 573(1969)
2. W. Liang, N. Goldsman, and I.Mayergoyz .VLSI Design 6, 251(1998)
3. A. Pirovano, A.L. Lacaita, and A. Spinelli. IEEE Trans. Electr. Dev. 49, 25(2002)
4. J.H. Rhew, Z. Ren, and M.S. Lundstrom. Solid State Electronics 46, 1899(2002)
5. K. Natori. J. Appl. Phys 76, 4879(1994)
6. K.Natori. Appl. Surf. Sc.254, 6194(2008)
7. J.P. Colinge, J.C. Alderman, W. Xiong, and C. Rinn Cleavelin. IEEE Trans. Electr. Dev. 53, 1131(2006)
8. A. Tsormpatsoglou, C.A. Dimitriadis, R. Clerc, G. Panakakis, and G.Ghibaudo. IEEE Trans. Electr. Dev. 55, 2623(2008)

CHAPTER 8

Post-Si FETs

8.1 INTRODUCTION

In the previous chapter we have dealt with the Si MOSFET, which is the mainstream device in contemporary electronics, but MOSFETs made out of other materials do exist and in fact may have superior transport properties compared to Si. One such material is GaAs and its alloys, and in general many of the III–V compound semiconductors and their alloys. We have seen in chapter 3 that GaAs is isotropic with an electron effective mass $m_n^* = 0.067 m_e$, which is significantly smaller than both of the two effective masses m_t^* and m_l^* of Si, making GaAs a much faster semiconductor than Si. Of course, the latter has other advantages which explain its use in the electronics industry for the past decades, namely, its availability in nature and the existence of a stable oxide, SiO_2, which leads to less expensive industrial processes compared to GaAs.

Because of their much better transport properties, the III–V materials constitute the building blocks of a new type of device called High-Electron-Mobility-Transistor (HEMT for short) with superior noise properties compared to Si MOSFETs. This device is suited ideally for high frequency telecommunication circuits. We analyze it first, because of its simplicity compared to the III–V MOSFETs, before we move on to the latter, which are also called quantum-well FETs (QWFETs for short). Finally, we analyze FETs which depart from the planar structure, such as the carbon nanotube FET (or CNTFET).

8.2 SIMPLE THEORY OF THE HEMT

a. **Device Description – Control by the Gate**

A simplified picture of a HEMT is shown in figure 8.1. On an undoped substrate of GaAs, a layer of $Al_{1-x}Ga_xAs$ is grown. The index x is usually near $x = 0.3$. The semiconductor $Al_{1-x}Ga_xAs$ is not uniformly doped but a very small layer, a few Angstroms thick, immediately adjacent to GaAs, is left undoped to reduce the scattering of the channel carriers, as will be explained later. On top of $Al_{1-ex}Ga_xAs$ a metal gate is deposited. The particular choice of III–V semiconductors shown in figure 8.1 is not unique and other pairs of III–V semiconductors with a difference in band-gap E_g are also used. This particular pair corresponds to one of the historically first analyzed

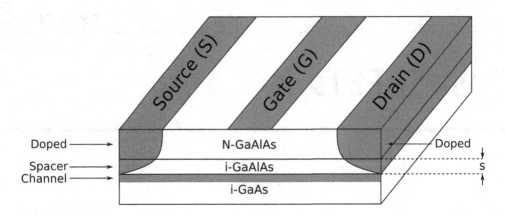

FIGURE 8.1   Layout of a III–V HEMT.

HEMT devices [1]. As with the Si MOSFET, where the Si/SiO$_2$ interface was the determining factor for the working of the device, the important physical phenomena occur in HEMTs at the interface of the two semiconductors. Therefore, we draw in a separate figure 8.2 the junction of the two semiconductors before and after contact.

In figure 8.2a we have on the left an N-type semiconductor (in this case AlGaAs) with a band-gap E_{g2} and a Fermi level E_{F2} near E_{C2}, its conduction band edge. On the right we have an intrinsic semiconductor (GaAs in this case) with a band-gap $E_{g1} < E_{g2}$ and a Fermi level E_{F1} at mid-gap. When contact of the semiconductors is made, the Fermi levels must be equalized, so electrons from AlGaAs will flow into GaAs. As with the PN and Schottky junctions, this flow of charge will destroy charge neutrality and a built-in potential will result, which will bend the band edges E_{C2} and E_{C1}. The resulting conduction band edges E_{C2} and E_{C1} are shown in figure 8.2b. Note that the spatial direction shown in figure 8.2b is along the depth of the device shown in figure 8.1. An inspection of figure 8.2b shows that a surface layer of electrons is created in the GaAs side of the junction (occupying the energy levels below E_F). If the well that has been formed is very narrow (of the order of a few nanometers), the allowed energy levels become quantized as shown. The electrons that have been assembled in this well form the channel of the HEMT shown in figure 8.1. We can now see the usefulness of the undoped layer in AlGaAs. If ionized donors were present there, they would have

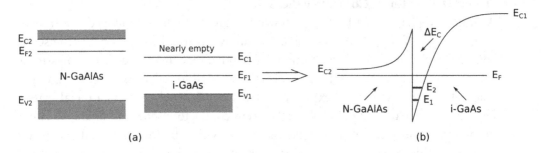

FIGURE 8.2   Band structure of (a) isolated N-GaAlAs and of isolated GaAs and (b) of the HEMT when the two pieces make contact with each other.

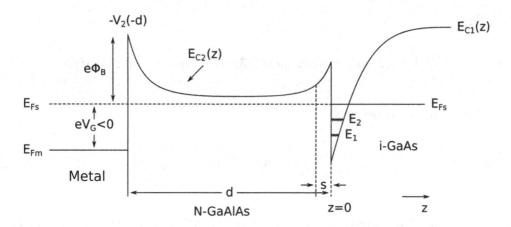

FIGURE 8.3   Band profile of the metal/N-GaAlAs/GaAs structure.

scattered the channel electrons strongly by their long-range Coulombic potential. In fact, the advantage of this device is that only the top layer is doped, so the electrons in the channel flow in an undoped material (GaAs), and therefore their mobility is enhanced—hence the name of the device. Furthermore, GaAs and its alloy GaAlAs have the same crystallographic structure, so their interface is smooth and hence the interfacial scattering is small compared to the Si/SiO$_2$ or Si/HfO$_2$ interfaces

When the metal gate is added to the picture, the corresponding band diagram including the Schottky junction below the gate is shown in figure 8.3. On the metal side, Φ_B stands for the Schottky barrier in volts. If a voltage V_G is applied to the gate which has to be negative so that electrons do not leak out of the gate, it can empty or fill the potential well in GaAs. To see how this is done, we have to solve the Poisson equation on the left and right side of the semiconductor interface. The argument is simple but rather long.

Let d be the thickness of the AlGaAs (its extent in the z-direction), and s the thickness of its undoped layer. Poisson's equation in the AlGaAs layer in one dimension reads

$$\frac{d^2 V_2}{dz^2} = \frac{-eN(z)}{\varepsilon_2} \qquad \text{8.1a}$$

where the point $z = 0$ is taken to be the interface of the two semiconductors, ε_2 is the dielectric constant of GaAlAs and

$$N(z) = N_D \text{ for } -d < z < -s, \; N(z) = 0 \text{ for } -s \leq z \leq 0 \qquad \text{8.1b}$$

The subscript 2 simply denotes that all quantities refer to the GaAlAs=semiconductor 2. Integrating once from an arbitrary negative point z to 0 we get

$$\left.\frac{dV_2}{dz'}\right|_{z'=0} - \left.\frac{dV_2}{dz'}\right|_{z'=z} = -\frac{e}{\varepsilon_2}\int_z^0 N_2(z')dz'$$

But

$$-\frac{dV_2}{dz'}\bigg|_{z'=0} \equiv \mathcal{E}_2\left(0^-\right) \equiv \text{electric field infinitesimally before the interface}$$

Hence we can write

$$\frac{dV_2}{dz'}\bigg|_{z'=z} = -\mathcal{E}_2\left(0^-\right) - \frac{e}{\varepsilon_2}\int_0^z N_2(z')dz' \qquad 8.2$$

where in 8.2 we have reversed the limit of integration in the RHS of the equation. Integrating a second time between –d and 0 we get

$$V_2(0) - V_2(-d) = -\mathcal{E}_2\left(0^-\right)[0-(-d)] - \frac{e}{\varepsilon_2}\int_{-d}^0\int_0^z N(z')dz'$$

and choosing our zero of energy to be $V_2(0)$ (i.e. $V_2(0)=0$) we get

$$V_2(-d) = d\mathcal{E}_2\left(0^-\right) - \frac{eN_D}{2\varepsilon_2}(d-s)^2 \qquad 8.3$$

where the 2nd term of the RHS of 8.3 arises from the double integration appearing in the last equation before 8.3. As we have repeatedly stressed, to obtain the potential energy of an electron corresponding to a potential V, one has to multiply by $(-e)$ so that the quantity $\left[-eV_2(-d)\right]$, shown in figure 8.3, is

$$-eV_2(-d) = \frac{eN_D}{2\varepsilon_2}(d-s)^2 - d\mathcal{E}_2\left(0^-\right) \qquad 8.4$$

Assuming now that the interface of $Al_xGa_{1-x}As/GaAs$ is free of charged defects, a reasonable assumption given that the two semiconductors have the same crystallographic structure, the normal component of the dielectric displacement vector $D^\perp$ is continuous and hence

$$D_1^\perp = D_2^\perp \Rightarrow$$

$$\Rightarrow \varepsilon_1 \mathcal{E}_1\left(0^+\right) = \varepsilon_2 \mathcal{E}_2\left(0^-\right) \qquad 8.5$$

where $\mathcal{E}_1\left(0^+\right)$ is the electric field infinitesimally inside semiconductor 1 (i.e.GaAs).

We do not have to solve the Poisson equation in GaAs to evaluate $\mathcal{E}_1\left(0^+\right)$. The electric field $\mathcal{E}_1\left(0^+\right)$ can be obtained easily by the use of Gauss's law. Imagine a rectangular box extending from infinitesimally to the right of the interface to deep inside GaAs. Electric field lines cross vertically only the plane coinciding with the interface and furthermore the electric field is zero far to the right. Hence if Gauss's law

$$\int D_1 \cdot ds = \text{charge enclosed}$$

is applied to this box, with A the cross-sectional area of the interface we get

$$\varepsilon_1 \mathcal{E}_1 (0^+) A = e \int n(z) dx dy dz \Rightarrow$$

$$\Rightarrow \varepsilon_1 \mathcal{E}_1 (0^+) = e \int_0^{z_{max}} n(z) dz = e n_s \qquad 8.6$$

where z_{max} is the maximum depth beyond which n_s is practically zero. In the above equation, $en(z)$ stands for the volume density (Cb/cm^3) and en_s is the surface density (Cb/cm^2).

From equations 8.4 to 8.6 we obtain

$$n_s = \frac{\varepsilon_2}{ed} \left[V_{ds} - V_2(-d) \right] \qquad 8.7a$$

where

$$V_{ds} = \frac{eN_D}{2\varepsilon_2} (d-s)^2 \qquad 8.7b$$

However, from figure 8.3 we have by summing and equating energy differences on the metal and the semiconductor sides (remember that $V_2(0) = 0$)

$$\Phi_B - V_G = V_2(-d) + \frac{\Delta E_C}{e} - \frac{E_F}{e} \qquad 8.8$$

Therefore using equation 8.8 to substitute for $V_2(-d)$

$$n_s = \frac{\varepsilon_2}{ed} \left(V_{ds} - \Phi_B - \frac{E_F}{e} + \frac{\Delta E_C}{e} + V_G \right) \qquad 8.9$$

The above equation is not a direct solution for the required surface density of electrons because the Fermi level E_F is not known. All other quantities are known. In fact, the Fermi level is a function of n_s, i.e. $E_F(n_s, T)$, so equation 8.9 could be solved self-consistently. This can be performed if the integral of the density of states with the Fermi–Dirac probability distribution is evaluated to obtain iteratively n_s and E_F. Fortunately, a reasonable approximation is available: the quantity E_F/e can be neglected compared to other quantities in 8.9 in most cases. We then get the required control relation of the channel by the gate as

$$n_s = \frac{\varepsilon_2}{ed}\left[V_G - V_{off}\right]$$ 8.10a

where

$$V_{off} = \Phi_B - V_{ds} - \frac{\Delta E_C}{e}$$ 8.10b

It should be obvious that the quantity V_{off} is the equivalent of V_{th} of the MOSFET: no current is possible below $V_G = V_{off}$, in this simplified model. However there is a subthreshold current, as it is called, which is due to diffusion which the following simplified model based on equation 8.10 does not include.

When a voltage V_D is applied between the drain and the source (it is henceforth understood that all voltages refer to the source), a longitudinal electric field $\mathcal{E}(x)$ is established along the channel. If this longitudinal field $\mathcal{E}(x)$ is small compared to the fields along the transverse z-direction, we can generalize 8.10 to

$$en_s(x) = \frac{\varepsilon_2}{d}\left[V_G - V_{ch}(x) - V_{off}\right]$$ 8.11

where $V_{ch}(x)$ is the potential drop along the channel up to the point x, caused by $\mathcal{E}(x)$. This approximation is the same we have used for the MOSFET (c.f. equation 7.12) and is usually called the gradual channel approximation. It actually replaces a more rigorous 2-dimensional solution of the Poisson equation by an effective equation of the current in terms of the electron density $n_s(x)$. Then, the current I_D will be

$$I_D = W\left[en_s(x)\right]v_{dr}(x)$$ 8.12

where W is the width of the device and $v_{dr}(x)$ is the drift velocity of the electrons. Note that to get the current from the current density, we only need to multiply by W because the other dimension of the cross-section area z_{max} is contained in n_s. From equation 8.12 it is obvious that if $v_{dr}(x)$ enters saturation near the drain then the corresponding $n_s(x)$ must become a constant, i.e. a residual channel will be created as invoked (but not proved) for the MOSFET.

If $\mathcal{E}_c$ is the critical value of $\mathcal{E}(x)$ at which saturation of velocity occurs at the value $v_{dr}(x) = v_{dr}^{sat}$ we can write

$$v_{dr}(x) = \mu\mathcal{E}(x), \text{ for } \mathcal{E}(x) \leq \mathcal{E}_c \tag{8.13a}$$

$$v_{dr}(x) = v_{dr}^{sat} = \mu\mathcal{E}_c, \text{ for } \mathcal{E}(x) \geq \mathcal{E}_c \tag{8.13b}$$

We initially assume that in the linear region of operation, equation 8.13a is valid for all points in the channel. Then, equation 8.12 is transformed into

$$I_D = \frac{\mu W \varepsilon_2}{d}\left[V_G - V_{ch}(x) - V_{off}\right]\frac{dV_{ch}}{dx} \tag{8.14}$$

In equations 8.12 to 8.14 we have neglected the minus signs for both the charge of an electron and the definition of an electric field in terms of the derivative of the potential, so I_D denotes the magnitude of current and is positive. Integrating and noting that the current I_D is independent of x we get

$$I_D\int_0^L dx = \int_{V_{ch}(0)}^{V_{ch}(L)} \frac{\mu W \varepsilon_2}{d}\left(V_G - V_{ch} - V_{off}\right)dV_{ch} \tag{8.15}$$

where L is the length of the channel. We can introduce the ohmic resistances R_S and R_D of the source and drain, respectively, so that we have for $V_{ch}(0)$ and $V_{ch}(L)$

$$V_{ch}(0) = R_S I_D \tag{8.16a}$$

$$V_{ch}(L) = V_D - R_D I_D \tag{8.16b}$$

From equation 8.15 we get

$$I_D = \frac{-\mu W \varepsilon_2}{2dL}\left[\left(V_G - V_{off} - V_{ch}\right)^2\right]_{V_{ch}(0)}^{V_{ch}(L)} =$$

$$= \frac{\mu W \varepsilon_2}{dL}\left(V_G - V_{off}\right)\left(V_{ch}(L) - V_{ch}(0)\right)$$

In deriving the above result we assumed the V_{ch} terms to be small compared to $2(V_G - V_{off})$ in the linear region. Using equations 8.16a and 8.16b we get

$$I_D = \frac{\mu W \varepsilon_2}{dL}\left(V_G - V_{off}\right)\left[V_D - I_D\left(R_S + R_D\right)\right] \tag{8.17}$$

Solving for I_D

$$\frac{V_D}{I_D} = R_S + R_D + \frac{Ld}{\mu W \varepsilon_2 \left(V_G - V_{off}\right)} \tag{8.18}$$

Equation 8.18 has an obvious physical interpretation: if the quotient V_D/I_D is to be interpreted as an effective resistance, the latter is the sum of the ohmic resistances plus a channel resistance which depends on V_G and goes to infinity as $V_G \to V_{off}$.

b. **Saturation Region**

As V_D is increased, so is $\mathcal{E}(x)$. There will therefore come a point when at the drain-end of the channel the electric field will become equal to $\mathcal{E}_c$. Then the device will enter the saturation regime in accordance with everything we have described for the MOSFET. Note however that the above statement refers to the general features of the saturation regime, but the algebraic expressions for either V_{off} or V_D^{sat}, (the saturation voltage) are not the same with the corresponding ones of the MOSFET. Performing the integration in equation 8.15 from $V_{ch}(0)$ to an arbitrary value $V_{ch}(x)$ where the point x lies before the drain-end of the channel we have

$$I_D x = \frac{\mu W \varepsilon_2}{d}\left[\left(V_G - V_{off}\right)\left(V_{ch}(x) - V_{ch}(0)\right) - \frac{V_{ch}^2(x)}{2} + \frac{V_{ch}^2(0)}{2}\right]$$

Defining

$$\frac{\mu W \varepsilon_2}{d} \equiv \frac{1}{C}$$

we can get a quadratic equation for $V_{ch}(x)$. We have

$$-\frac{1}{2}V_{ch}^2(x) + \left(V_G - V_{off}\right)V_{ch}(x) - \left(V_G - V_{off}\right)V_{ch}(0) + \frac{1}{2}V_{ch}^2(0) - CI_D x = 0 \tag{8.19}$$

After some trivial algebraic manipulations the determinant of the above equation becomes

$$\Delta = 4\left[\left[V_G - V_{off} - V_{ch}(0)\right]^2 - CI_D x\right]$$

Hence the physically acceptable solution of equation 8.19 is

$$V_{ch}(x) = V_G - V_{off} - \sqrt{\left[V_G - V_{off} - V_{ch}(0)\right]^2 - CI_D x} \tag{8.20}$$

As noted in the beginning of this subsection, pinch-off will occur first at the drain-end of the channel. Hence

$$\left. \left| \frac{dV_{ch}(x)}{dx} \right| \right|_{x=L} = \mathcal{E}_c$$

The term pinch-off above is used within the context of the previous chapter on the MOSFET, i.e. the depth of the channel never becomes zero, but a very small portion is left open, etc. Differentiating equation 8.20 and assigning

$$I_D \rightarrow I_D^{sat} \text{ at } \mathcal{E} \geq \mathcal{E}_c$$

we get

$$CI_D^{sat} \frac{1}{\sqrt{\left[V_G - V_{off} - V_{ch}(0)\right]^2 - CI_D^{sat}L}} = \mathcal{E}_c \Rightarrow$$

$$\Rightarrow C^2 \left(I_D^{sat}\right)^2 - \mathcal{E}_C^2 \left(\left[V_G - V_{off} - V_{ch}(0)\right]^2 - CI_D^{sat}L\right) = 0 \qquad 8.21$$

We now restore back in equation 8.21 the values of C and use 8.13b to substitute the saturation velocity v^{sat} wherever the product $\mu\mathcal{E}_C$ appears. We obtain

$$\left(I_D^{sat}\right)^2 + \frac{2L\mathcal{E}_C W\varepsilon_2 v_{dr}^{sat}}{d} I_D^{sat} - \left(\frac{W\varepsilon_2 v_{dr}^{sat}}{d}\right)^2 \left[V_G - V_{off} - V_{ch}(0)\right]^2 = 0 \qquad 8.22$$

The appropriate solution of this quadratic equation is, making the substitution $V_{ch}(0) = I_D^{sat}R_S$,

$$I_D^{sat} = \frac{W\varepsilon_2 v_{dr}^{sat}}{d} \left[\sqrt{\mathcal{E}_C^2 L^2 + \left(V_G - V_{off} - I_D^{sat}R_S\right)^2} - \mathcal{E}_C L\right] \qquad 8.23$$

If $\mathcal{E}_C^2 L^2 \ll \left(V_G - V_{off} - I_D^{sat}R_S\right)^2$, i.e. if we have a short channel HEMT, we can simplify equation 8.23 to

$$I_D^{sat} = \frac{W\varepsilon_2 v_{dr}^{sat}}{d} \left(V_G - V_{off} - I_D^{sat}R_S - \mathcal{E}_C L\right) \qquad 8.24$$

The saturation current of the long channel HEMT obeys a different equation. It is left as an exercise for the student to obtain it.

c. **AC Analysis**

All the machinery we have developed for the AC analysis of the MOSFET equally holds for the HEMT and can be taken over here. In particular, equation 7.60 and the corresponding methodology holds for any three-terminal device with a negligible gate current. So, following equation 7.60 and the approximation 7.61, we obtain the same small-signal equivalent circuit. Following the same path as in equations 7.69 to 7.71 we obtain, as before, the cut-off frequency f_T

$$f_T = \frac{g_m}{2\pi C_{GS}} \qquad \text{8.25a}$$

where C_{GS} stands, as in the MOSFET, for the capacitance linking the gate to the source.

Obtaining an algebraic expression for C_{GS} is not as easy as it was for the MOSFET. A very crude estimate of this capacitance can be obtained by omitting the charges in the top semiconductor. Then we can write

$$C_{GS} = \frac{\varepsilon_2}{d + \Delta d} \qquad \text{8.25b}$$

where ε_2 is the dielectric constant of the top semiconductor and Δd is the distance of the maximum of the channel charge density from the interface. But this is not a reliable expression to calculate f_T. However, the value of the f_T can be obtained and be related to the length of the channel L by a semi-quantitative argument as follows. This argument is in many respects more general than the above formula. Imagine the electrons crossing the channel. During their flight imagine a sinusoidal change $v_{gs}(t)$ in the external circuit (we are using the notation of section 7.6). If the electrons take a time t_{fl} to reach from the source to the drain (i.e. cross the channel) and during that time $v_{gs}(t)$ has undergone many cycles, then the electrons will never reach the drain. In other words, the excitations in the external circuit must change in time more slowly than the time required to traverse the channel. We can safely assume that modern FETs operate at saturated velocities, so this t_{fl} time is

$$t_{fl} = \frac{v_{dr}^{sat}}{L} \qquad \text{8.26a}$$

and hence

$$f_T = \frac{v_{dr}^{sat}}{2\pi L} \qquad \text{8.26b}$$

8.3 ADVANCED THEORY OF THE HEMT

To obtain a value for V_{off} in equation 8.10 without resorting to numerical computations we had to omit the term E_F/e in equation 8.9. If we wanted to calculate E_F we would need to solve Schrödinger's equation along the depth z of the device, obtain the eigenvalues and then obtain n_s. In fact our analysis of the HEMT so far was essentially a 1-dimensional one: we solved the Poisson equation in 1 dimension along the depth z of the device and used a simple drift equation along the x direction. From our discussion of the MOSFET in chapter 7 we saw how inefficient these 1-dimensional models are. The presence of both V_G and V_D imply a vertical (to the gate) field and a parallel field, respectively, making any transistor problem essentially a 2-dimensional problem, especially at short gate lengths. The machinery we have developed for the MOSFET can be applied equally well to the HEMT. In fact it becomes even simpler in the case of the HEMT because GaAs and several of its alloys are direct band-gap materials having the respective conduction band-edge E_C at $k = 0$, so there is no valley dependency.

We summarize below the set of differential equations, specifically for the HEMT, that are necessary to produce a more accurate (though numerical) simulation than the simple algebraic model that we have given so far.

Poisson's Equation

$$\nabla \varepsilon \nabla V = -\rho = -\left(N_D^+ - n\right) \qquad 8.27$$

In equation 8.27 the dielectric constant is put between the 3-dimensional derivatives because HEMT structures include many layers of III–V materials with different dielectric constants.

The 2-Dimensional Schroedinger's Equation (Effective mass form).

$$\left[-\frac{\hbar^2}{2}\frac{\partial}{\partial x}\frac{1}{m^*}\frac{\partial}{\partial x} - \frac{\hbar^2}{2}\frac{\partial}{\partial z}\frac{1}{m^*}\frac{\partial}{\partial z} + V(x,z)\right]\Psi_i(x,z) = E_i\Psi_i(x,z) \qquad 8.28$$

In equation 8.28 a unique effective mass m^* is used on the one hand but on the other hand the form of the equation valid for interfaces was used (see equation 5.40) because again many layers of different effective masses are necessary for the analysis of the HEMT. The Schroedinger equation, contrary to the Poisson equation, is not solved in the entire domain of the device but only in a subdomain, called quantum box, which includes the channel layer extending well into the source and drain regions, just as in the MOSFET.

The Continuity Equation

$$\nabla J_n = 0 \qquad 8.29$$

In equation 8.29 the recombination term R_n has been omitted since it is small in unipolar devices and especially in HEMTs. Equation 8.29 can give us an equation for the Fermi level

E_F as follows: the expression for the drift-diffusion current for electrons, equation 7.24, can be written as (see problem 3.2 at the end of chapter 3)

$$J_n = n\mu_n \nabla E_F \qquad 8.30$$

If we substitute 8.30 in 8.29 we get

$$n\mu_n \nabla^2 E_F + \mu_n \nabla E_F \nabla n = 0 \qquad 8.31$$

Given the eigenfunctions $\Psi(x,z)$ and the Fermi level E_F, the quantum charge density $n_Q(x,z)$ can be calculated in analogy with equation 7.30 for the MOSFET.

$$n_Q(x,z) = e \sum_i |\Psi_i(x,z)|^2 \int f_0(E,E_F) g_{1D}(E) dE \qquad 8.32$$

where f_0 is the Fermi- Dirac occupation function. A notable difference, however, with the MOSFET is the following. The depth of the well in the oxide/Si interface is determined by the band-edge offset ΔE_C of Si with respect to the oxide which is $\Delta E_C = 3.9 eV$. The corresponding quantity in HEMTs is the band-edge offset ΔE_C of very similar III–V alloys (e.g. InGaAs and AlGaAs), resulting in a ΔE_C of a few tenths of an eV, i.e. much smaller than the Si/SiO$_2$ interface. This low depth of the well holding the 2-dimensional electron gas results in some eigenstates being above the top of the barrier E_{top}, but below E_F, i.e. some occupied eigenstates are not true 2-dimensional states, see figure 8.4. The situation is erected by summing in 8.32 only eigenstates with energies $E_i < E_{top}$. For the remaining eigenstates we simply use the corresponding 3-dimensional formula, treating them as a classical semiconductor electron density, i.e.

$$n_{cl} = N_C F_{1/2} \left(\frac{E_F - E_{top}}{KT} \right) \qquad 8.33$$

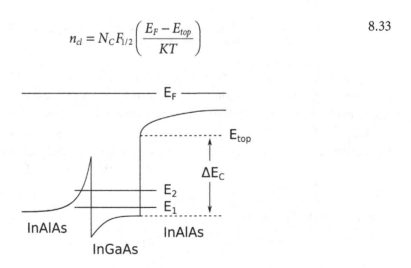

FIGURE 8.4   Definition of the quantity E_{top} in a quantum well above which the energy eigenvalues E_1, E_2 etc may be considered as 3-dimensional.

Then, the total electron density n is given by

$$n = n_Q + n_{cl} \qquad\qquad 8.34$$

A flow chart that solves the HEMT equations is given below.

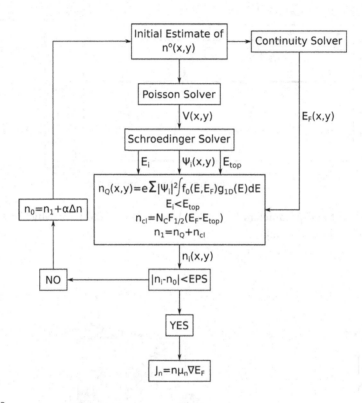

FLOWCHART 3

In the above flow chart, EPS is the accuracy with which we want to calculate n, the electron density. Note that in the return box to the Poisson equation the electron density n is updated not by the new value, but by the old value n_0 plus a fraction α of the change Δn in n. To ensure convergence that fraction must be as low as 1%. The workload can be significantly reduced for long-channel HEMTs ($L > 100nm$) if the Schroedinger equation is left out. Then the updated charge density is $n_1 = n_{cl}$ with $E_{top} = E_C$ and $n_Q = 0$. Such a calculation is called classical whereas the one including the Schroedinger equation is frequently called quantum, although we have reserved the term quantum in this book for a calculation that uses a quantum-mechanical expression for the current.

Some typical results for the HEMT structure are given below. Figure 8.5 gives the particular HEMT structure for which the author has performed calculations for several values of the gate length [2]. In this particular structure, an InGaAs channel is sandwiched between two InAlAs barriers. The channel density for $L = 100nm$ and $V_{GS} = V_{DS} = 0$ is given in Figure 8.6. Note 1) the high electron density below the source and drain regions which

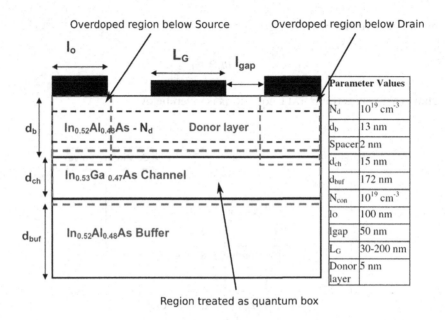

FIGURE 8.5 Layout and parameters for an InAlAs/InGaAs HEMT for which calculations have been performed by the author [2].

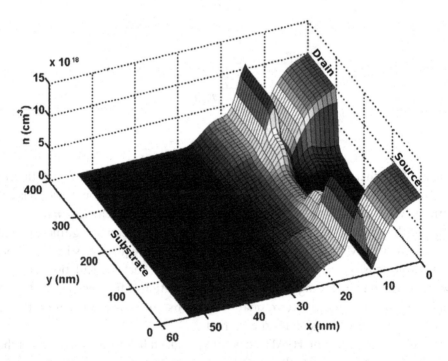

FIGURE 8.6 A 2-dimensional view of the electron density across the entire structure of the device of figure 8.5. The gate and drain voltages are $V_{GS} = V_{DS} = 0$. Note the channel electron density, having just been formed, the high electron density below the source and drain regions, which are overdoped, and the depleted InAlAs region below the gate.

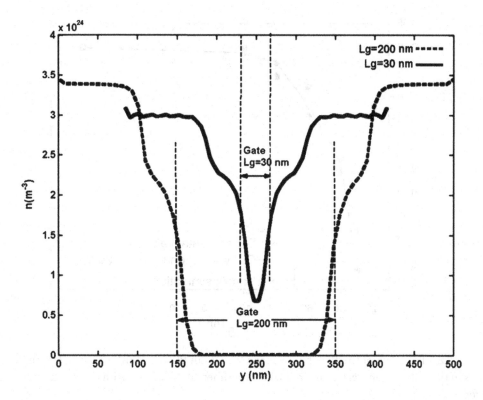

FIGURE 8.7   A 1-dimensional view of the electron density of a nanometric and a long-channel HEMT: whereas the long HEMT is below its threshold the nanometric one is above its threshold.

are overdoped, 2) the depleted region below the gate in the InAlAs layer, and 3) the channel region having just been formed in the InGaAs layer. Note also that in order to achieve high accuracy a large depth of the buffer onto which the channel rests had to be included in the calculations.

A 1-dimensional cut along the channel length at the middle point of the channel depth is shown in figure 8.7 for two values of the channel length, $L = 200nm$ and $L = 30nm$ at $V_{GS} = V_{DS} = 0$. It can be easily seen that whereas the $L = 200nm$ HEMT is below threshold, the $L = 30nm$ HEMT is above. Obviously these two HEMTs have different threshold voltages and this brings us to the first of the short channel effects mentioned in the previous chapter, that is the lowering of the threshold voltage with decreasing L. This is shown for the HEMT under investigation in figure 8.8. The symbol ΔV_T instead of ΔV_{off} is used on the vertical axis. A very low drain voltage, $V_{DS} = 0.05$, is applied. It can be seen that the change in the threshold voltage is substantial, of the order of half an eV. The effect of the change in V_{off} on the saturation current is shown in figure 8.9 where the calculated I_D–V_D characteristics are plotted for three values of the gate length L, i.e. at $L = 30, 60, 100nm$. We observe that irrespective of whether a classical or a quantum (in the sense of including the Schroedinger equation) calculation is performed, there is a monotonic increase of the saturation current when L is decreased from $L = 100nm$ to $L = 30nm$. The above phenomena occur in all HEMTs, not only in the one presented above.

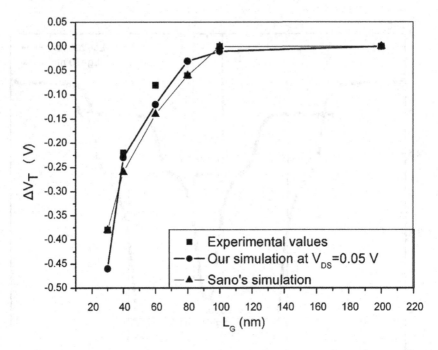

FIGURE 8.8   Lowering of the threshold voltage V_{off} (V_T in the diagram) with gate length L_G. The values of this calculation were compared with experimental values and with other simulations; see [2] for more details.

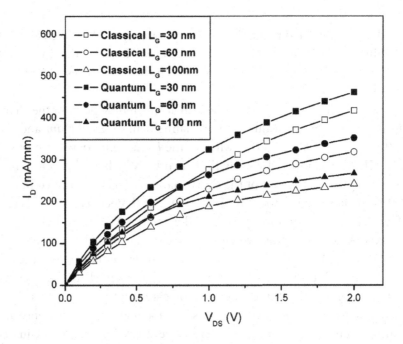

FIGURE 8.9   Increase of the current I_D with drain voltage V_{DS} as the gate length L_G decreases in both classical and quantum calculations; see text for details.

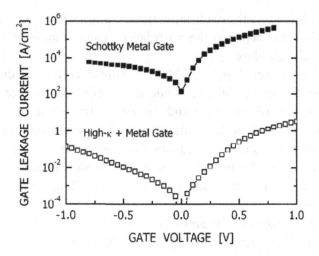

FIGURE 8.10   Comparison of the leakage current of HEMT and III–V MOSFETS. Figure reproduced from R. Chau et al Microelectronic Eng. 80, 1 (2005)

8.4 THE III–V MOSFET

The previously discussed HEMT takes advantage of the high mobility of the III–V semiconductors but suffers from a disadvantage compared to the Si MOSFET: the Schottky gate with which it operates exhibits a high leakage current as shown in figure 8.10. The remedy for this disadvantage is the introduction of a dielectric under the gate, thus making a complete analog to the Si MOSFET. However, this proved to be a formidable task. Existing dielectrics HfO_2 and Al_2O_3 are not easily deposited without creating surface defects which scatter the electrons and slow down their response. Fortunately, ways of passivating the defects have been devised and nowadays III–V MOSFETs with both HfO_2 and Al_2O_3 under the gate exist. Figure 8.11 portrays two versions of III–V

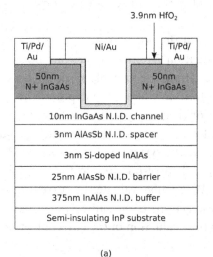

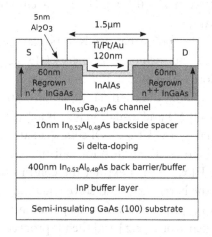

(a)   (b)

FIGURE 8.11   Typical III-V MOSFET devices.

MOSFETs, one with HfO_2 as the oxide layer and the other with Al_2O_3. Observe the complexity of the devices, in particular the many layers required for stability and to contain the current in the channel layer.

The theoretical tools to analyze III–V MOSFETs are essentially no different than what has been presented for both Si MOSFETs and III–V HEMTs. However, due to the complexity of the devices on the one hand and the nanometric length of the channel on the other hand a combination of both quantum (i.e. Landauer) methods and classical equations together with the Schroedinger equation are needed to fully describe the device. Furthermore, some additional modifications of the classical equations are necessary to fully account for the effects of the nanometric size of the channel that we have discussed in the previous chapter.

In the closing paragraphs of the previous chapter we emphasized that the saturation regime of the Si MOSFET is not dictated by the saturation velocity occurring at the pinch-off point but instead by the injection velocity over the top of the barrier that is near the source-end of the channel. The latter has nothing to do with the term v_{dr}^{sat} that we have used in the analysis of the long Si MOSFET. Guided by these considerations, we can allow the saturation velocity entering the mobility expression in the continuity equation to attain higher values than those usually quoted in the literature, essentially treating this quantity as a parameter.

One may ask why we need the continuity equation when the Natori formula will directly give the current? The answer is that to use the Natori formula, equation 7.48, requires the eigenvalues $\varepsilon_{n_z}\left(x_{top}\right)$ of equation 7.40 and this, in turn, requires the construction of the corresponding potential energy U'. If the latter is to be calculated self-consistently the system of equations 8.27–8.34 needs to be solved. We note that if the depth of the quantum well were of the order of $2-3eV$ as in Si MOSFET we could assume that all electrons reside in the lowest sublevel of the well. However, as explained in the case of the HEMT, the conduction band-offset for III–V semiconductors is of the order of a few tenths of an eV (between $0.2-0.4eV$). Hence the employment of the full set of equations 8.27–8.34 is necessary.

A further point that needs to be clarified is that the potential energy $V(x,z)$ of equation 8.28 and the potential energy $U(x,z)$ of equation 7.40 in chapter 7 are not the same. Equation 7.40 is essentially a 1-dimensional equation, the variable x simply denotes the position along the channel. Equation 7.40 is solved only along the depth of the channel. On the other hand, equation 8.27 is a 2-dimensional equation and so is the corresponding potential energy $V(x,z)$. Of course, at any point x, $U(x,z)$ is the corresponding column of $V(x,z)$. To use the Natori formalism we first solve the system of equations 8.27–8.34 and obtain $V(x,z)$. Then extract the column $U\left(x_{top},z\right)$ corresponding to the position of the top of the barrier x_{top}, and solve equation 7.40. We finally evaluate the current using 7.49. Accordingly, flowchart 4 corresponding to these actions is the same as flowchart 3 until convergence of $V(x,z)$ and then it proceeds to the evaluation of the current by the Natori method.

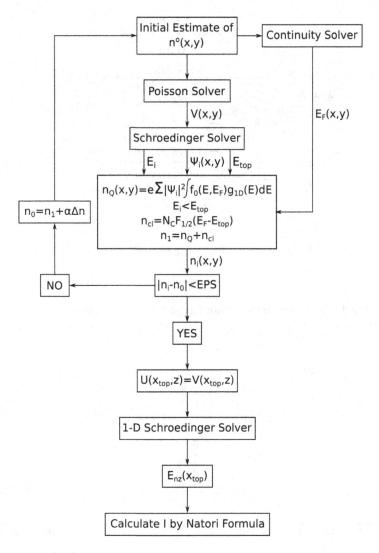

FLOWCHART 4

In the above flowchart one should not confuse the symbol E_{top} which refers to the top energy value of the 2-dimensional potential well with $\varepsilon_{n_z}\left(x_{top}\right)$ which are the eigenvalues of the 1-dimensional well along z located at x_{top} which is used in the Natori formula.

Some calculations by the author [3] for a typical device portrayed in figure 8.12 which show the influence of the many layers in III–V as opposed to Si MOSFETs, are shown in Figure 8.13. The figure gives the longitudinal current density $J_x\left(z\right)$ as a function of the depth z of the device for several values of the gate voltage V_G. The channel layer ($\mathrm{In}_{0.53}\mathrm{Ga}_{0.47}\mathrm{As}$) is located between $z = 70nm$ and $z = 80nm$. We observe that the current density J_x below threshold ($V_{off} < 0$) is located near the doped area and gradually, as the gate voltage is increased above threshold, it shifts in the channel layer. This shifting of the current density

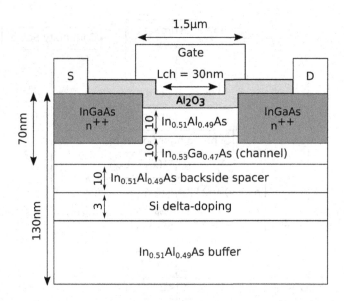

FIGURE 8.12   The device for which the author has performed calculations.

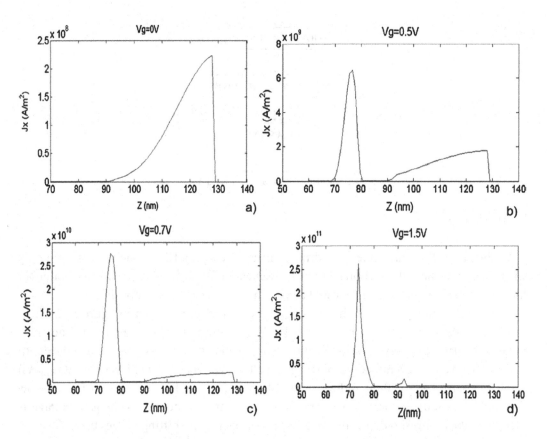

FIGURE 8.13   Variation with depth of the longitudinal current density as the gate voltage increases: the subthreshold current occurs below the channel and gradually moves to the channel area.

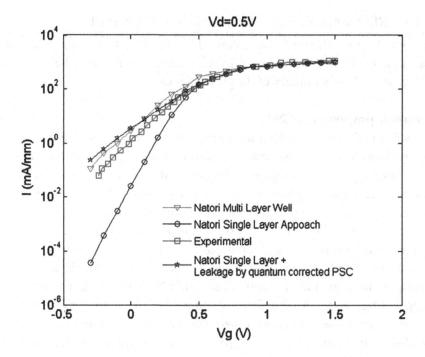

FIGURE 8.14 Current as a function of gate voltage for the device shown in figure 8.12. When the quantum well in which the current is calculated by the Natori method is identical to the channel, the calculated subthreshold current deviates from the experimental values. When the quantum well is enlarged to include lower layers, agreement with experiments is restored; see text for more details.

can also be seen if one looks at the current as a function of V_G. This is shown in figure 8.14 together with the experimental results. The current is calculated in two different ways: the circles represent a calculation where the quantum box (the domain where the Schroedinger equation is solved) coincides with exactly the channel (InGaAs) area while the triangles represent a calculation where the quantum box includes the layers above and below the channel. In both cases the Natori formula is used. Consequently the 1 dimensional well, where the $\varepsilon_{n_z}(x_{top})$ are calculated, includes in the latter case the barrier layers also. The former calculation coincides with Natori's original assumption that the wavefunctions are contained mainly within the channel layer with only an exponentially decreasing tale outside this layer. It can be seen that the calculation with the larger quantum box (the one including the barriers) is very near the experimental results, whereas the calculation with the quantum box containing only the channel layer misses all the current that exists below the channel at sub-threshold. Above threshold the two calculations agree because all the current is contained within the $In_{0.53}Ga_{0.47}As$ layer. Then if the quantum box is bigger than the $In_{0.53}Ga_{0.47}As$ layer makes no difference. A third calculation, denoted by asterisks, is shown in which the current is calculated in the InGaAs layer by the Natori method and in the layers below by the Poisson-Schroedinger-Continuity (PSC) equations. This also agrees with experiment. Therefore great care should be exercised when using the Natori formalism to many layer III–V MOSFETs.

8.5 THE CARBON NANOTUBE FET, CNFET, OR CNTFET

We now come to the ultimate 1-dimensional FET, the FET that has a carbon nanotube (CNT) as a channel and is surrounded by an oxide from all sides. Before tackling the CNTFET we give a brief summary of the properties of CNTs.

a. **Electronic properties of CNTs**

A CNT is produced by rolling up a graphene sheet (i.e. a single layer of atoms of the graphite structure) along one of its 2-dimensional lattice vectors $R = n_1 a_1 + n_2 a_2$ as shown for example in figure 8.15 where a (6,4) CNT is depicted. This process will produce an infinite tube of nanometric diameter. We then expect, given the 1-dimensionality of the CNT, to obtain a 1-dimensional Brillouin zone. How does this come about? The argument is the following—the planar graphene sheet has a planar Brillouin zone, the hexagon depicted in figure 8.16a. The unit vectors b_1, b_2 in k space are related to the unit lattice vectors a_i by the relation $b_j \cdot a_i = 2\pi\delta_{ij}$ (see the corresponding equation in chapter 2, equation 2.18). When the rolling up of a CNT along a lattice vector R is performed and a CNT of radius $r = |R|/2\pi$ is produced, then the atom at the origin and at the tip of R become one and the same atom. Their wavefunctions must be equal and this is guaranteed (using Bloch's theorem) by the relation

$$k \cdot R = 2\pi m \qquad\qquad 8.35$$

where m is an integer. This will discretize the hexagonal Brillouin zone of graphene into a series of line segments as shown in figure 8.16b.

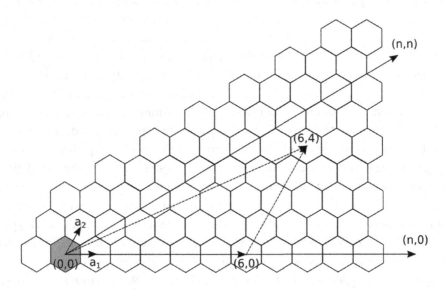

FIGURE 8.15 The 2-dimensional lattice of grapheme showing a typical lattice (n,m) vector along which if a graphene is rolled it will produce an (n,m) CNT.

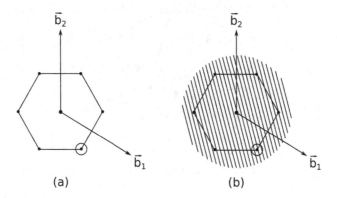

FIGURE 8.16 (a) The hexagonal Brillouin zone of graphene and (b) its transformation into a series of line segments when a CNT is produced by the rolling of a graphene sheet.

A simple mathematical proof is as follows. We can analyze k into components along the folding vector R and perpendicular to the folding vector R. Denote these by k_x and k_y, respectively. Then we have that $R_y=0$ and equation 8.35 takes the form

$$k_x R_x + k_y R_y = 2\pi m \Rightarrow k_x R_x = 2\pi m \Rightarrow$$

$$\Rightarrow k_x = \frac{2\pi m}{|R|} \qquad\qquad 8.36$$

which means that k_y may assume any value whereas k_x is discretized in multiples of $2\pi/|R|$. Hence, the allowed k vectors are line segments perpendicular to the folding vector R. They are contained within the graphene hexagonal Brillouin zone, and are separated by a successive distance of $2\pi/|R|$. As the magnitude $|R|$ of R tends to large values we recover the 2-dimensional character of the graphene but for $|R|$ of the order of nanometers the number of line segments inside the hexagon is finite and each can be considered as a sub-band of the (n_1,n_2) nanotube. The k_y values are also discrete but as the length of the CNT increases the k_y tend to a continuum.

Carbon nanotubes can be either metallic or semiconducting. The analysis of the previous paragraph allows us to deduce a very simple rule as to whether a (n_1,n_2) CNT is metallic or semiconducting. The vertices of the hexagonal Brillouin zone (BZ) are the wavevectors at the Fermi energy, i.e. the k_F. So if the line segments corresponding to the allowed k values cross (or contain) the vertices of the hexagon there are allowed states at the Fermi energy and the CNT is metallic. If not, they are semiconducting. Figure 8.17 is a magnification of the allowed states near the bottom right corner of the hexagonal BZ of figure 8.16b. The broken line shows the direction of the folding vector R, so the line segments containing the allowed states are perpendicular to it.

Consider an arbitrary state given by k and denote by $\Delta k_{\parallel}$ and $\Delta k_{\perp}$ the components of $|k-k_F|$ along the line segment and perpendicular to it. Clearly if $\Delta k_{\perp}$ is zero the

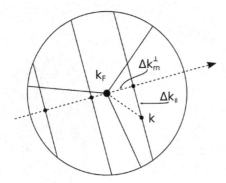

FIGURE 8.17   Magnification of figure 8.16b near one of the corners of the hexagonal BZ.

particular line segment contains the particular vertex and the CNT is metallic. The component $\Delta k_\perp$ equals the projection of $|k - k_F|$ onto R. Then

$$\Delta k_\perp = \left(k - k_F\right) \cdot \frac{R}{|R|} = \frac{k \cdot R - k_F \cdot R}{|R|} \qquad \qquad 8.37$$

The particular vertex we have encircled is given by

$$k_F = \frac{b_1 - b_2}{3}$$

From 8.36 we then get

$$\Delta k_\perp = \frac{2\pi m}{|R|} - \frac{b_1 \cdot R - b_2 \cdot R}{3|R|} =$$

$$= \frac{1}{|R|} \left[2\pi m - \frac{b_1 \cdot \left(n_1 a_1 + n_2 a_2\right) - b_2 \cdot \left(n_1 a_1 + n_2 a\right)}{3} \right] =$$

$$= \frac{1}{|R|} \left[2\pi m - 2\pi \frac{\left(n_1 - n_2\right)}{3} \right] \qquad \qquad 8.38$$

In the last, before the last line of 8.38 we have used the relation between the unit direct and reciprocal lattice vectors discussed above (equation 2.18 of chapter 2)

$$a_i \cdot b_j = 2\pi \delta_{ij} \qquad \qquad 2.18$$

Hence we get

$$\Delta k_\perp = \frac{2\pi}{3|R|} \left(3m - n_1 + n_2\right) \qquad \qquad 8.39$$

So if $(n_1 - n_2)$ is a multiple of 3, $\Delta k_\perp = 0$ and the CNT is metallic, otherwise it is semi-conducting with a bandgap E_g that depends on the folding vector. It should be noted that the result is independent of the particular vertex chosen for the proof.

The band structure of the CNTs can be approximated by the band structure of graphene (i.e. curvature effects are omitted) by an application of the LCAO method that we have presented in detail in chapter 2. This is left as an exercise to the student, see problems at the end of the chapter. The result is

$$E = E_0 \pm t \sqrt{1 + 4\cos^2\left(\frac{k_y a}{2}\right) + 4\cos\left(\frac{k_x \sqrt{3}a}{2}\right)\cos\left(\frac{k_y a}{2}\right)} \qquad 8.40$$

where E_0 and t are the diagonal and off-diagonal elements respectively of the p orbitals of graphene and a is the interatomic distance (see the instructions for the corresponding problem). The density of states of the semiconducting (10,0) CNT is shown in figure 8.18a and that of the metallic (9,0) CNT in figure 8.18b. Equation 8.40 takes

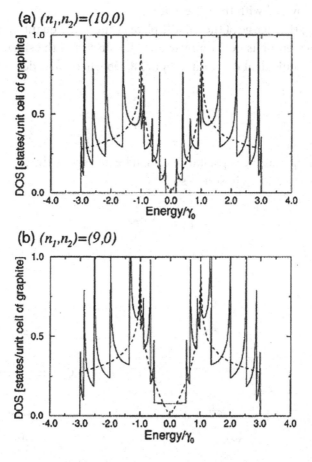

FIGURE 8.18 (a) Density of states of a semiconducting (10,0) CNT and (b) of a metallic (9,0) CNT. Calculations by R. Saito et al, Appl. Phys. Lett. 60, 2204 (1992).

into account the discretization in one of the directions of the 2-dimensional **k** vector, but it does not take into account the curvature of the CNT. As a general trend, it can be stated that in some metallic (according to the above rule) CNTs curvature may open up a small gap if their diameter is small. We will not deal with this subject, instead assuming that all CNTs that form a FET channel are semiconducting.

b. **The concept of a quantum capacitance**

The concept of a quantum capacitance was introduced by S. Luryi in 1998 [4] as a correction to the gate capacitance of a MOSFET. However, this notion has become important in connection to the functioning and modeling of the CNTFET so we will give a brief introduction to the concept. We have seen in the previous chapter that as the gate voltage V_G is increased above V_{th} the conduction band of Si is populated and an almost surface charge Q_n is created at the Si-SiO$_2$ (Si-HfO$_2$ nowadays) interface. This charge screens the electric field lines emanating from the gate and leads to the stabilization of the width of the space charge layer in Si. The situation is depicted in a more general form in the following figure 8.19a of a three-plate capacitor. If the middle plate is indeed metallic any increase of V_G will increase the electric field only in the space covered with the dielectric ε_1.

However, if the charge of the middle plate is due to the quantum well of a semiconductor, the screening is not complete and the physics of the situation may be represented by the equivalent diagram of figure 8.19b where C_Q is defined as

$$C_Q = \frac{dQ}{dE_F} \qquad 8.41$$

Then, assuming that the capacitance C_2 can be neglected, the capacitance C_G seen by the gate is just C_1 and C_Q in series, so

$$C_G = \frac{C_1 C_Q}{C_1 + C_Q} \qquad 8.42$$

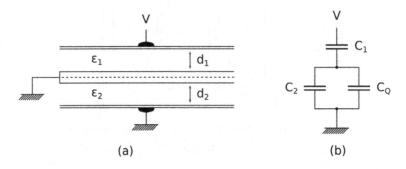

(a)   (b)

FIGURE 8.19 (a) Two capacitors in series in which the middle plate is not perfectly conducting and (b) its equivalent circuit.

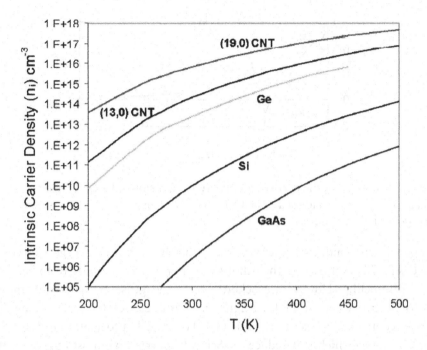

FIGURE 8.20   Comparison of the intrinsic carrier density n_i of several semiconductors. Figure reproduced from D. Akiwande et al [5].

c. DC model of CNTFET

Carbon nanotubes may be doped producing N- and P-type semiconductors but it is not necessary to dope CNTs to produce a FET. The intrinsic carrier concentration of $(n,0)$ CNTs is as high as that of semiconductors with common doping, see figure 8.20. Obviously the high values of $n_i(T)$ for CNTs displayed in figure 8.20 stem from their very low volume. Doped CNTs however are used as the source and drain contacts of a FET. Indeed, the two most common forms of CNTFETs are shown in figure 8.21. In figure 8.21a the source and drain contacts are formed by overdoped CNTs and an intrinsic CNT constitutes the channel. On the other hand, in 8.21b the source and drain are formed by metals. A brief discussion is necessary of these metal contacts before proceeding to a model of the CNTFET

We have more than once emphasized that a well-designed MOSFET needs ohmic contacts at the source and drain in order to function properly. This is usually achieved by overdoping the semiconductor which reduces the length of the Schottky barrier at

FIGURE 8.21   Two forms of CNT-FETs.

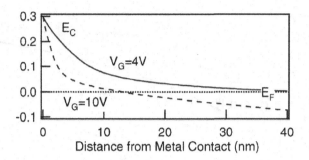

FIGURE 8.22 Reduction of the barrier length between a metal and a CNT by the application of a gate voltage in the initial stage of a CNT-FET research when ohmic contacts were not available. Figure reproduced from Avouris et al [6].

the metal–semiconductor junction. Unfortunately, in the initial stages of research into the CNTFET development, the Schottky barrier lengths were far from being permeable by tunnelling. However the Schottky barrier lengths were reduced by the application of the gate voltage V_G, [6], see figure 8.22. This led to a device which was Schottky barrier controlled instead of a usual FET. Fortunately procedures and materials are available today which may reduce the Schottky barrier to almost zero. We now come to the evaluation of the current I_D. We follow closely the analysis of Akinwande et al [7].

The band diagrams of the CNTFETs which are shown in figure 8.21 are given in figure 8.23. These band diagrams assume a Schottky barrier of almost zero. Note that because the Fermi level of the intrinsic CNT is almost at the middle of its band gap and that of the heavily doped CNT very near E_C, the difference $E_C - E_F = E_{g/2}$ where E_F is the Fermi level of the channel. The dispersion relations of the subbands (those arising from the discretization of k_x) of both the conduction and the valence band are shown in figure 8.24. Only the first two sub-bands are shown for each band. The effect of the application of a gate voltage $V_G > 0$, as shown in figure 8.24, is to shift the electron sub-bands down, This will populate the conduction band predominantly as can be seen from an inspection of figure 8.23. We may safely assume that the number of holes is negligible. The amount by which the conduction sub-bands move down is just the surface potential φ_s (see equations 7.1a and 7.1b of chapter 7 and figure 7.5b).

Then the current I_D will be given by the Landauer formula just as we did for the MOSFET, see equations 7.43–7.47 of chapter 7. Furthermore, we take account of

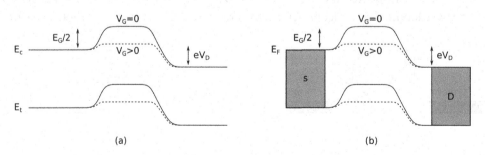

FIGURE 8.23 Band diagrams of the CNTFETs shown in figure 8.21.

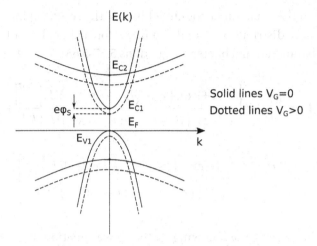

FIGURE 8.24   Dispersion relations (E-k) of the subbands of a CNT.

contributions from only the two lowest conduction subbands. Denoting the equilibrium Fermi–Dirac distribution by $f(E_F, E)$ and by E_{C1} and E_{C2} the subband minima, see figure 8.24, we have

$$I_D = e \int_{E_{C1} - e\varphi_s}^{\infty} M(E)g(E)v(E)\left[f(E_{FS}, E) - f(E_{FD}, E)\right]T(E)dE +$$

$$+e \int_{E_{C2} - e\varphi_s}^{\infty} M(E)g(E)v(E)\left[f(E_{FS}, E) - f(E_{FD}, E)\right]T(E)dE \qquad 8.43$$

Note that in equation 8.43 above it was no longer necessary to distinguish between the direction of transport x and other directions. There is only one direction, that along which transport takes place which is normal to the barrier. As with 7.43 $g(E)$, $v(E)$ stand for the 1-dimensional density of states and the velocity at energy E respectively. The factor $M(E)$ is the subband degeneracy and is equal to $M(E) = 2$ for carbon nanotubes. Assuming ballistic transport and performing the simplifications we have used in going from 7.43 to 7.47 we get (choosing the Fermi level at the source $E_{FS} = E_F$)

$$I_D = \frac{4e}{h} \int_{E_{C1} - e\varphi_s}^{\infty} \left[f(E_F, E) - f(E_F - eV_D, E)\right]dE +$$

$$+\frac{4e}{h} \int_{E_{C2} - e\varphi_s}^{\infty} \left[f(E_F, E) - f(E_F - eV_D, E)\right]dE \qquad 8.44$$

where h in the denominator in 8.44 is Plank's constant (not h bar).

Contrary to the situation encountered in 7.47 the resulting integrals involve only the Fermi–Dirac distribution f and can be performed analytically (see the formula for the supply function in chapter 5, equations 5.67-5.68). We get

$$I_D = \frac{4ekT}{h} \ln\left[\frac{1+exp\left[\left(E_F -(E_{C1} -e\varphi_s)\right)/KT\right]}{1+exp\left[\left(E_F -eV_D -(E_{C1} -e\varphi_s)\right)/KT\right]} \right] +$$

$$+\frac{4ekT}{h} \ln\left[\frac{1+exp\left[\left(E_F -(E_{C2} -e\varphi_s)\right)/KT\right]}{1+exp\left[\left(E_F -eV_D -(E_{C2} -e\varphi_s)\right)/KT\right]} \right] \qquad 8.45$$

It is worthwhile noting the following: 1) the above equation does not explicitly show the V_G dependence of I_D, this is hidden in the surface potential φ_s, 2) the current from drain to source is only important in the linear region, it can be dropped in the saturation region.

To complete the model the gate voltage dependence of φ_s must be evaluated. The latter can be obtained by either a self-consistent numerical calculation (Poisson–Schroedinger type) or can be obtained by a compact (empirical) model. It has been shown that φ_s can be given by the following analytic compact model [7],

$$\varphi_s = \frac{C_{ox}}{C_{ox} + \dfrac{C_Q(\varphi_s)}{2}} V_G \qquad 8.46$$

where C_Q is the quantum conductance discussed in the previous subsection.

The above model reproduces the $(I-V)$ curves of CNTFETs only moderately. Its primary deficiency is the omission of optical phonon scattering near the drain end of the channel [7]. In fact this is a common characteristic of most nano-FET transistors, which brings us to our original question in chapter 5 of where the power is dissipated. It is dissipated near the drain end of the channel and in the thermodynamic reservoirs where electrons are thermalized.

PROBLEMS

8.1 Find the saturation current of a long channel (length) HEMT using the same methodology as in the text.

8.2 The small-signal equivalent circuit of the HEMT in the text was assumed to be topologically the same as that of the MOSFET. Both, however, were simplified versions. If the resistances R_S and R_D were to be included in the small-signal equivalent circuit, how this would be modified?

8.3 Use the LCAO theory of chapter 2 to prove the dispersion relation (equation 8.39) of carbon nanotubes neglecting curvature effects (i.e. using grapheme as a model of a CNT). You may ignore the s orbitals as well as the p_x, p_y orbitals (the ones lying on the plane of the carbon atoms.

8.4 A model of CNT is shown in the following diagram (The model was produced by researchers of the Universities of Purdue, Cornell and Stanford). Prove that

$$V^{CNT} = \frac{C_{ox}}{C_{ox}+C_D}V_G + \frac{C_D}{C_{ox}+C_D}V_D + \frac{Q^{CNT}}{C_{ox}+C_D}.$$

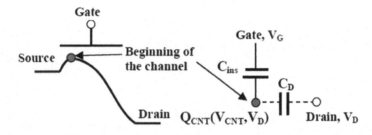

REFERENCES

1. D. Delagebeaudeuf and N.T. Linh. IEEE Trans.Elect.Dev 29, 955(1982)
2. G. Krokidis, J.P. Xanthakis, and N. Uzunoglu. Solid State Electronics 52, 625(2008)
3. A. Giliand J.P. Xanthakis Solid State Electronics 136, 55(2017)
4. S. Luryi Appl.Phys.Lett. 52, 501(1988)
5. D. Akinwande, Y.Nishi and H.S. Philip Wong IEEE IEDM (2007)
6. S. Heinze, J. Tersoff, R. Martel, V. Derycke, J. Appenzeller and Ph, Avouris Phys. Rev. Lett. 89, 106801 (2002)
7. D. Akiwande, J. Liang, S. Chong, Y.Nishi, and H.S.P. Wong J.Appl.Phys. 104, 125514(2008)

Appendix A: Further Development of Quantum Mechanics, Angular Momentum, and Spin of the Electron

Before embarking on an analysis of angular momentum and spin of the electron, we collect together here our main findings of Quantum Theory. In section 1.2 and 1.3 we learned that the energies of the electrons are given by the eigenvalues of the hamiltonian H and the latter is constructed by substituting for the momentum the operator $-i\hbar\nabla$ in the classical expression for the energy. The eigenfunctions $\Psi(r)$ of the hamiltonean H, are then used to deduce the average position and momentum of an electron in a potential energy field. This concept was further generalized in section 1.4. Just as the expression for the classical energy of an electron was turned into an operator, the hamiltonian operator H, the expression for any classical quantity Q is turned into an operator $\hat{Q}$ in Quantum Mechanics by exactly the same rules as for the hamiltonean H, i.e. 1) $p \rightarrow -i\hbar\nabla$ and 2) leaving the position vector r as r. Then the possible measured values of that physical quantity are the eigenvalues q_n of that operator. Mathematically we should write

$$\hat{Q}|q_n\rangle = Q_n|q_n\rangle \qquad\qquad \text{A1}$$

where $|q_n\rangle$ are the eigenvectors of $\hat{Q}$. The average value of repeated measurements is then

$$\bar{Q} = \frac{\int \Psi^*(r)\hat{Q}\Psi(r)dxdydz}{\int \Psi^*(r)\Psi(r)dxdydz} \qquad\qquad \text{A2}$$

Note that this is a generalization of equations 1.11 and 1.12.

The physical thinking behind equation A1 is that the act of measurement and the result of measurement are not the same thing in Quantum Mechanics. The measurement of variable A and then of variable B will not necessarily give the same result if it is performed in the opposite order, first measurement of B, then A. To account for this, we must in some way have, (not always),

$$AB - BA \equiv [A,B] \neq 0 \qquad\qquad \text{A3}$$

This can only be achieved with operators. The act of measurement in Quantum Mechanics is represented by Hermitian operators which have the property of having real eigenvalues and for which $A_{m,n} = (A_{m,n})^*$ (where the star denotes complex conjugate). When A3 is indeed different from zero, we say that A and B do not commute and $[A,B]$ is called the commutator of A,B. However, in some cases, if $AB = BA$, this has a significant physical consequence: it means that the two physical variables–observables can be measured simultaneously. In this case, the two operators have common eigenfunctions. An example of this pair is the momentum and the energy in a zero potential field (where the energy is just the kinetic one only).

The angular momentum L is defined classically by

$$L = r \times p \qquad\qquad \text{A4}$$

The components of L are classically and then quantum mechanically as follows:

$$L_x = yp_z - zp_y \rightarrow -i\hbar\left(y\frac{\partial}{\partial z} - z\frac{\partial}{\partial y} \right) \qquad\qquad \text{A5a}$$

$$L_y = zp_x - xp_z \rightarrow -i\hbar\left(z\frac{\partial}{\partial x} - x\frac{\partial}{\partial z} \right) \qquad\qquad \text{A5b}$$

$$L_z = xp_y - yp_x \rightarrow -i\hbar\left(x\frac{\partial}{\partial y} - y\frac{\partial}{\partial x} \right) \qquad\qquad \text{A5c}$$

It is easier to work in the spherical coordinates (r,θ,φ) given by

$$x = r\sin\theta\cos\varphi \qquad\qquad \text{A6a}$$

$$y = r\sin\theta\sin\varphi \qquad\qquad \text{A6b}$$

$$z = r\cos\theta \qquad\qquad \text{A6c}$$

Elementary calculus manipulations give

$$L_x = i\hbar\left(\sin\varphi\frac{\partial}{\partial\theta} - \cot\theta\cos\varphi\frac{\partial}{\partial\varphi} \right) \qquad\qquad \text{A7a}$$

$$L_y = i\hbar \left(-\cos\varphi \frac{\partial}{\partial\theta} + \cot\theta\sin\varphi \frac{\partial}{\partial\varphi} \right) \qquad \text{A7b}$$

$$L_z = -i\hbar \frac{\partial}{\partial\varphi} \qquad \text{A7c}$$

It is easy to see that

$$[\hat{L}_x, \hat{L}_z] \neq 0 \quad \text{and} \quad [\hat{L}_y, \hat{L}_z] \neq 0 \qquad \text{A8}$$

However we find that L_z commutes with the magnitude of L which is given by

$$\hat{L}^2 = \hat{L}_x^{\,2} + \hat{L}_y^{\,2} + \hat{L}_z^{\,2} =$$

$$= -\hbar^2 \left[\frac{1}{\sin\theta} \frac{\partial}{\partial\theta} \left(\sin\theta \frac{\partial}{\partial\theta} \right) + \frac{1}{\sin^2\theta} \frac{\partial^2}{\partial\varphi^2} \right] \qquad \text{A9}$$

This can easily be deduced by inspection of equation A9. Therefore L^2 and L_z can be measured simultaneously and hence they must have common eigenfunctions. These eigenfunctions are the spherical harmonics $Y_{l,m}(\theta,\varphi)$ that we have encountered in the theory of the Hydrogen atom. A rather lengthy purely mathematical argument shows that the eigenvalue equations for L^2 and L_z take the form

$$\hat{L}^2 Y_{l,m}(\theta,\varphi) = \hbar^2 l(l+1) Y_{l,m}(\theta,\varphi) \qquad \text{A10}$$

$$\hat{L}_z Y_{l,m}(\theta,\varphi) = \hbar m_l Y_{l,m}(\theta,\varphi) \qquad \text{A11}$$

$$\text{with } l = 0,1,2,\ldots\ldots \qquad \text{A12}$$

and

$$m_l = 0,\pm1,\pm2,\ldots,\pm l \qquad \text{A13}$$

Therefore the eigenvalues of L^2 are $\hbar^2 l(l+1)$ and the eigenvalues of L_z are $\hbar m_l$. The integers l and m_l are identical to the quantum numbers we have encountered in the electron states of the Hydrogen atom. We conclude that the hydrogenic electrons with a given l state are in a state of constant (magnitude of) angular momentum.

The above theory is not complete. The famous Stern–Gerlach experiment, which we describe immediately below, has shown that the electron possesses in addition to the orbital angular momentum (due to its motion around the nucleus) an inherent "spin" angular momentum as if it were spinning around itself. The experiment briefly is as follows: A beam of Silver (Ag) atoms is produced in an oven and passed through a magnetic

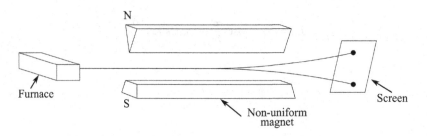

FIGURE A1   Apparatus of the Stern–Gerlach experiment.

field pointing in the z direction, see figure A1. The Ag atoms have an atomic number $Z = 47$ with 46 electrons in a closed inner shell with no angular momentum (the sum of the orbital momenta along the z direction cancels out) and the 47th electron is in a $5s^1$ orbital which also has zero "orbital" angular momentum (because $l = 0$ in this state).

Classically a circulating charge (the electron) possesses a magnetic moment perpendicular to its plane of circulation and of magnitude $\mu = IA$ where I is the current produced by the charge and A the area enclosed by the circulating charge. The magnetic field of the magnet used in the experiment is non-uniform so that a force F_z will be exerted on the outer electrons of Ag given by

$$F_z = -\frac{\partial U}{\partial z} \qquad\qquad\text{A14}$$

where

$$U = -\mu \cdot B \qquad\qquad\text{A15}$$

Since the magnetic moment μ is proportional to the angular momentum that creates it and the z-component of the latter is quantized (according to A11) we should see as many spots, after the beam has passed through the magnet, as the number of possible values of L_z.

Two spots of Silver atoms are observed on the screen. This is however in contradiction to equations A12 and A13. If $l = 0$ there should have been one spot and if $l \geq 1$ there should have been three or more. The conclusion of the Stern–Gerlach experiment is that the electron possesses an angular momentum additional to the orbital angular momentum called spin angular momentum which is given the symbol $\hat{S}$ because it is a different quantity from $\hat{L}$. For the eigenvalues S_z we should have in analogy with equations A11 and A13

$$S_z = \hbar m_s \text{ , with } m_s = \pm\frac{1}{2}$$

Therefore m_s plays the role of a fourth (fractional) quantum number in addition to $n, l,$ and m_l that we have encountered in chapter 1. Each state defined by (n, l, m_l) can be occupied by two electrons, one with $S_z = \hbar / 2$ and another with $S_z = -\hbar / 2$. We usually refer to them as spin up and spin down electrons. Note that the Pauli principle is not violated because no two electrons have the same (four) quantum numbers.

Appendix B: Lattice Vibrations

In chapter 2 we treated the crystalline lattice as stationary. This is of course not true. In chapter 3 we discussed the acceleration of electrons by an external electric field and the subsequent scattering of them by their interaction with the vibrating lattice. In this Appendix we give the elements of a quantitative theory of lattice vibrations. As usual we will start with elementary notions and subsequently build on these.

Consider first a periodic linear chain with 1 atom per unit cell. This is shown in figure B1. The forces on each atom can be considered as Hooke-like $(F = -kx)$ with the bonds between atoms acting as springs. Then the force on each atom will be proportional to the displacements, $u_l(t)$, where l denotes lattice site, of the nearby atoms from their equilibrium position. We assume only nearest neighbour interactions. Then using Newton's third law and denoting by M the mass of each atom and by k the spring constant of the corresponding force

$$M \frac{d^2 u_l}{dt^2} = k\left[u_{l+1} - u_l - (u_l - u_{l-1})\right] =$$

$$= -k\left[2u_l - u_{l+1} - u_{l-1}\right] \tag{B1}$$

But the u_l must satisfy Bloch's theorem since they constitute waves in a periodic medium. Then we must have

$$u_{l+1} = e^{qa} u_l \tag{B2}$$

where q is a wavenumber and a is the distance between nearest neighbour. We get

$$M \frac{d^2 u_l}{dt^2} = -k\left(2u_l - e^{iqa} u_l - e^{-iqa} u_l\right) =$$

$$= -2k(1 - cosqa)u_l \tag{B3}$$

This is a typical harmonic oscillator equation.

Assuming a harmonic time dependence of the form

$$u_l(t) = u_l(0)e^{i\omega t} \tag{B4}$$

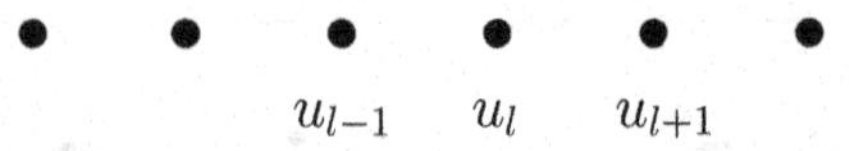

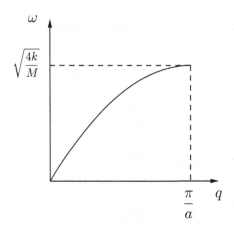

FIGURE B1   Notation for the positions of the atoms of a monoatomic chain.

we arrive at the frequencies of oscillation

$$\omega^2 = \frac{2k}{M}(1-cosqa) = \frac{4k}{M}sin^2\left(\frac{qa}{2}\right) \qquad \text{B5}$$

where in B5 a simple trigonometric formula has been used. At very small ω, B5 may be further simplified as follows

$$\omega \approx 0 \Rightarrow \omega \approx qa\sqrt{\frac{k}{M}} \qquad \text{B6}$$

The graph of $\omega(q)$ is shown in figure B2. The range of vibrational frequencies shown in figure B2 is said to constitute the acoustic mode of frequencies of a linear monoatomic chain because it resembles the waves found in a continuous elastic medium. As we will see, additional modes of vibrations exist in a linear chain if it contains more than one atom per unit cell.

Consider then a linear periodic chain with two atoms per unit cell. Let the displacements of the two types of atoms of the unit cell be denoted u_i^1 and u_i^2 and their respective masses M_1 and M_2. For M_1 we will have assuming a unique spring constant k between M_1 and M_2, see figure B3.

$$M_1\frac{d^2u_i^1}{dt^2} = k\left[u_i^2 - u_i^1 - \left(u_i^1 - u_{i-1}^2\right)\right] =$$

$$= k\left[u_i^2 + u_{i-1}^2 - 2u_i^1\right] \qquad \text{B7}$$

FIGURE B2   Frequencies of lattice vibrations of a monoatomic chain. These frequencies constitute the acoustic branch.

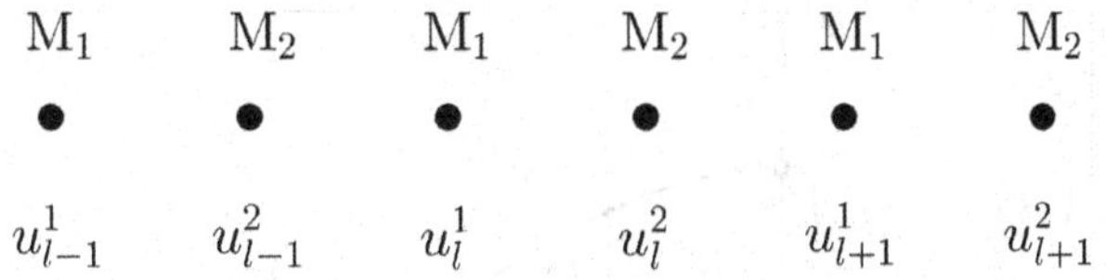

$$M_1 \quad\quad M_2 \quad\quad M_1 \quad\quad M_2 \quad\quad M_1 \quad\quad M_2$$

$$\bullet \quad\quad \bullet \quad\quad \bullet \quad\quad \bullet \quad\quad \bullet \quad\quad \bullet$$

$$u_{l-1}^1 \quad\quad u_{l-1}^2 \quad\quad u_l^1 \quad\quad u_l^2 \quad\quad u_{l+1}^1 \quad\quad u_{l+1}^2$$

FIGURE B3   Notation for the position of the atoms of a diatomic chain.

and for M_2

$$M_2 \frac{d^2 u_l^2}{dt^2} = k \left[u_{l+1}^1 + u_l^1 - 2u_l^2 \right] \tag{B8}$$

Assuming now 1) a harmonic variation for both u_l^1 and u_l^2 exactly as in equation B4 and 2) a Bloch relation for u_l^1 and u_l^2 separately i.e.

$$u_{l+1}^1 = e^{qa} u_l^1$$

and

$$u_{l-1}^2 = e^{-qa} u_l^2$$

we transform equations B7 and B8 into

$$-\omega^2 M_1 u_l^1 = k u_l^2 \left(1 + e^{-qa}\right) - 2k u_l^1 \tag{B9}$$

$$-\omega^2 M_2 u_l^2 = k u_l^1 \left(1 + e^{qa}\right) - 2k u_l^2 \tag{B10}$$

For the system of equations B9 and B10 to have a solution for u_l^1 and u_l^2 their determinant must be zero. Hence

$$\begin{vmatrix} 2k - M_1 \omega^2 & -k\left(1 + e^{-qa}\right) \\ -k\left(1 + e^{qa}\right) & 2k - M_2 \omega^2 \end{vmatrix} = 0 \tag{B11}$$

Equation B11 gives a second degree algebraic equation for ω^2 for each q. The form of the solution is shown in figure B4. It can be seen that in addition to the acoustic mode of vibration (the one going linearly to zero as $q \to 0$) a second mode of vibration has appeared which is called optical mode of vibration occurring at higher frequencies and hence energies. The reason for the name optical mode comes from the fact that the range of frequencies involved are very near those of the optical spectrum of the electromagnetic radiation. It is exactly the interaction with these type of vibrations that leads to the constancy of the electron drift velocity at high electric fields.

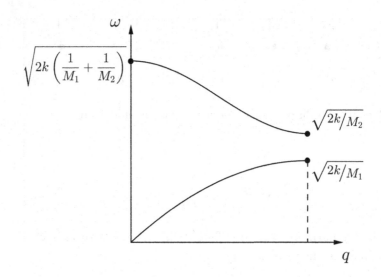

FIGURE B4   The dispersion relation ω(q) of both acoustic and optical frequencies.

Real solids of course cannot be described by linear chains and we therefore need a 3-dimensional generalizations of B7 and B8. We denote by the vectors u_l^i the displacements in 3 dimensions of atom type i in unit cell l (vector). The cartesian components of this vector we call u_l^{ij}, i.e. j denotes direction in space. The most efficient way of performing this generalization is to assume that the u_l^{ij} are small enough so that the potential energy V of the lattice can be expanded in a Taylor series in terms of the u_l^{ij} measured from their equilibrium zero values. We will have

$$V = V_0 + \sum_{i,j,l} u_l^{ij} \left.\frac{\partial V}{\partial u_l^{ij}}\right|_0 +$$

$$+ \sum_{\substack{i,j,l \\ i',j',l'}} u_l^{ij} u_{l'}^{i'j'} \left.\frac{\partial^2 V}{\partial u_l^{ij} \partial u_{l'}^{i'j'}}\right|_0 \qquad \text{B12}$$

where in B12 the symbol $|_0$ means evaluated at the equilibrium values of zero displacements. Now the constant term V_0 will give zero force and the first derivatives will be zero at the equilibrium position of the atoms, i.e. that of zero u_l^{ij}. Hence the only term which will contribute to the interatomic forces will be the last quadratic term. Denoting the second derivatives

$$\frac{\partial^2 V}{\partial u_l^{ij} \partial u_{l'}^{i'j'}} = K_{ll'}^{ij,i'j'}$$

the equation of motion can be written as

$$M_i \frac{d^2 u_l^{ij}}{dt^2} = -\sum_{i',j',l'} K_{ll'}^{ij,i'j'} u_{l'}^{i'j'}$$ B13

or in matrix form

$$M_i \frac{d^2}{dt^2} \mathbf{u}_l^i = -\sum_{i',l'} K_{ll'}^{i,i'} \mathbf{u}_{l'}^{i'}$$ B14

As there are many indices entering B13 and B14 we offer the following physical explanation of B14: each term on the RHS of B14 is the force exerted on atom type i in the lth unit cell due to the vector displacement $\mathbf{u}_{l'}^i$ of atom type i'in the l'th cell. The sum of all the forces on atom type i in the lth cell gives the corresponding acceleration. Obviously the number of equations in B14 is tremendous but Bloch's theorem can be used to reduce them to a small number. This reduction is accomplished by the following intermediate steps.

1. The force constants can only depend on the relative distance between unit cells

$$K_{ll'}^{i,i'} = K^{ii'}(l' - l)$$

2. As already noted the displacements $\mathbf{u}_l^i$ from waves in a crystalline solid, so they must obey Bloch's theorem

$$\mathbf{u}_{l+m}^i = e^{iq \cdot m} \mathbf{u}_l^i$$

Substituting the above equations in B14 eliminates the summations over l and we get

$$M_i \frac{d^2}{dt^2} \mathbf{u}_0^i = -\sum_{i'} \left[\sum_m K^{ii'} e^{iq \cdot m} \right] \mathbf{u}_0^{i'}$$ B15

Equations B15 form the generalization of B7 and B8. The 0th lattice site is of course only nominal so that u_0^i denotes the displacements of any unit cell. Equations B15, however, are much smaller in number compared to B14: The number of atoms per unit cell = N is certainly a small number and so is the number of interactions between the atoms; on the other hand if m separates atoms more than second nearest neighbours the interactions can almost always be ignored. As a general rule the number of vibrational modes is $3N$, and of them 3 are acoustic ones so that the optical ones are $3N - 3$.

The above presented theory is a classical one. One should have constructed a hamiltonian and quantized it. This is beyond the scope of the present book but if such a quantization is performed then the frequencies of oscillation would be quantized and the energies would occur in multiples of $\hbar\omega(q)$. However, the distinction between acoustic and optical modes would remain and also the fact that it is the optical modes that absorb energy from the electrons and lead to the stabilization of drift velocity.

Appendix C: Impurity States in Semiconductors

One of the historically first applications of the effective mass equation was the calculation of the donor and acceptor states, E_D and E_A, in semiconductors, see section 2.4. When a Si atom is replaced by an As atom which has 5 electrons in its outer shell (as opposed to 4 in silicon) a perturbing potential energy is created which, to a first level of approximation, can be taken to be that of a point charge in a dielectric medium, that is

$$V(r) = \frac{-e^2}{4\pi\varepsilon_0\varepsilon_r r} \qquad \text{C1}$$

The same argument can be repeated for a column VI of the periodic table donor in GaAs.

In order to get a feeling of the magnitude involved in E_D and E_A we can, to a first level of approximation, replace the effective mass tensor m_{ii}^* by an average effective mass m^* in silicon (Si) which has a degenerate conduction band. No such approximation is needed for direct band semiconductors as, for example GaAs. Then if the effective mass equation for electrons is used to calculate E_D, with the expression C1 substituted for the perturbing potential, the problem becomes identical with that of the hydrogen atom with the difference that m goes to m^* and ε_r appears in the denominator of the potential energy.

We can immediately deduce the eigenvalues (depth below E_c) from the hydrogen case, equation 1.40,

$$E_D^n = -\frac{m^* e^4}{8\varepsilon_0^2\varepsilon_r^2\hbar^2 n^2} \qquad \text{C2}$$

We can easily see that a) because of the minus sign the levels lie below the conduction band-edge and b) because of the $\varepsilon_r^2 (\approx 100)$ in the denominator of C2 and $m^* < m$ in the numerator, the $n = 1$ level will be of the order of tens of meV. One may question the use of the effective mass equation for such a rapidly varying potential as that of equation C1. However the justification for the use of the effective mass equation lies in the extent of the wavefunction obtained for the impurity states: we deduce, using equation 1.38, that the "equivalent" Bohr radius would be enlarged by a factor of $m\varepsilon_r/m^* (\approx 100)$ so that the region $r \approx 0$ when $V(r)$ of equation C1 is varying rapidly will not play a role. Similar arguments hold for the acceptor states.

Appendix D: Direct and Indirect Band-Gap and Optical Transitions

In chapter 2 we have seen that, as far as the band-gap is concerned, there are two kinds of semiconductors, those in which the conduction band minimum E_C and the valence band maximum E_V are at the same $\boldsymbol{k}$ point (called direct band-gap semiconductors), and those in which E_C and E_V are at different $\boldsymbol{k}$ points (called indirect band-gap semiconductors). A typical example of the first kind is GaAs, in which both E_V and E_C are at $\boldsymbol{k} = 0$, and of the latter is Si in which E_V lies at $\boldsymbol{k} = 0$ but E_C is six-fold degenerate and lies somewhere along such axes as (100). The two types of $E(\boldsymbol{k})$ curves are shown schematically in figure D1.

This difference in band-gaps is of a paramount importance for their optical properties. In short, indirect band-gap semiconductors cannot exhibit optical emission whereas direct semiconductors do and are the basic materials for the manufacturing of LASERs and LEDs. Both the interaction of an incident photon with an electron and the emission of a photon by an excited electron can be considered as a scattering process in which the energy and the momentum are conserved.

For a relaxation process in which an excited electron at E_C relaxes to E_V by emission of a photon of energy $\hbar\omega$ we have

$$E_C = E_V + \hbar\omega \text{ (conservation of energy)} \qquad \text{D1}$$

$$\boldsymbol{p}(E_C) = \boldsymbol{p}(E_V) + \boldsymbol{p}(photon) \text{ (conservation of momentum)} \qquad \text{D2}$$

where $\boldsymbol{p}(E_C)$ is the momentum of the excited electron at E_C, $\boldsymbol{p}(E_V)$ is the momentum of the electron at E_V after relaxation and $\boldsymbol{p}(\text{photon})$ the momentum of the emitted photon. Equation D2 may be written

$$\hbar\boldsymbol{k}(E_C) - \hbar\boldsymbol{k}(E_V) = \boldsymbol{p}(\text{photon}) \qquad \text{D3}$$

But photons carry momentum: according to the De Broglie hypothesis, electrons and photons are symmetrical, both have momentum p and wavelength λ, related to each other by $p = 2\pi\hbar/\lambda$

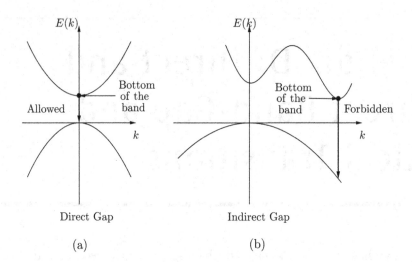

FIGURE D1   Schematic E (**k**) band diagrams showing (a) a direct and (b) an indirect band-gap semiconductor. Allowed and forbidden transitions are also shown.

(in 1 dimension). It is worth pointing out that even classically the electromagnetic field has momentum. Therefore

$$|\Delta\boldsymbol{k}|(\text{electron}) = \frac{2\pi}{\lambda} \qquad\qquad \text{D4}$$

where λ is the wavelength of the photon.

However, the wavevectors of electrons are of the order π/a where a is the interatomic distance, a few Angstroms, whereas the wavelength of visible light is of the order of $10^2\,nm \approx 10^3\,Å$. Hence the change $|\Delta\boldsymbol{k}|$ of an electron must be negligible. Obviously in our argument the implicit assumption that the electron has not taken up momentum from the lattice has been made. This restriction (that $|\Delta\boldsymbol{k}| \approx 0$) leads to what is called a vertical transition and has important consequences for the emission process in a semiconductor. If an electron finds itself in the conduction band minimum E_C of a direct band-gap semiconductor it can relax by a vertical transition to the valence band maximum E_V because empty states are available there (assuming the semiconductor is doped by acceptors), see

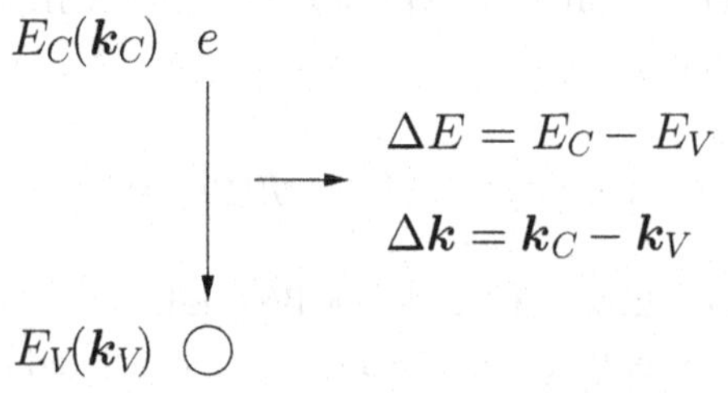

$$
\begin{aligned}
&E_C(\boldsymbol{k}_C) \quad e \\[2pt]
&\qquad\Big| \qquad\quad \Delta E = E_C - E_V \\
&\qquad\quad\longrightarrow \\
&\qquad\Big| \qquad\quad \Delta k = \boldsymbol{k}_C - \boldsymbol{k}_V \\[2pt]
&\qquad\downarrow \\
&E_V(\boldsymbol{k}_V) \quad \bigcirc
\end{aligned}
$$

FIGURE D2   Energy and momentum conservation during an electron relaxation from the conduction to the valence band.

the arrows in figure D1a. But if the excited electron is in an indirect semiconductor it cannot relax by a vertical transition because it will find itself in occupied states, see the arrow in figure D1b. Note that this transition is not allowed because of the combined effect of two laws, conservation of momentum and the Pauli's principle.

Note that the reverse transition (i.e. the absorption of a photon by the excitation of an electron) in an indirect semiconductor is not forbidden: an electron can go from an occupied state in the valence band to an empty state in the conduction band by a vertical transition. However, we have emphasized the emission process because this process forms the basis for the operation of LEDs and LASERs. In these devices the electrons find themselves in excited states because of the application of an external voltage.

Appendix E: Proof of the Field Emission Formula

We begin with the expression for the current density J

$$J = \frac{emkT}{2\pi^2 \hbar^3} T(E_F) \int_{E_b}^{\infty} exp\left[d_F(E_x - E_F)\right] \cdot ln\left[1 + exp\left(\frac{E_F - E_x}{kT}\right)\right] dE_x$$

We perform a change of variable $u = d_F(E_x - E_F)$, and we define $c \equiv d_F kT$. We can assume that $c < 1$ and that the bottom of the conduction band is sufficiently deep, i.e. $E_b \ll -1/d_F$. We get

$$J = \frac{emkT}{2\pi^2 \hbar^3 d_F} T(E_F) \int_{-\infty}^{\infty} e^u ln\left[1 + e^{\frac{-u}{c}}\right] du$$

We shall prove that the above integral, $I = \int_{-\infty}^{\infty} e^u ln\left[1 + e^{\frac{-u}{c}}\right] du$, is equal to $\pi/sin(\pi c)$.

We perform an integration by parts

$$I = \int_{-\infty}^{\infty} e^u ln\left[1 + e^{\frac{-u}{c}}\right] du = \left[e^u ln\left(1 + e^{\frac{-u}{c}}\right)\right]_{-\infty}^{\infty} + \frac{1}{c}\int_{-\infty}^{\infty} \frac{e^u}{e^{\frac{u}{c}} + 1} du$$

Now assuming $c < 1$, the first term is 0. For the second we perform the change of variable $v = e^{\frac{u}{c}}$ and we get

$$I = \int_{0}^{\infty} \frac{v^{c-1}}{v + 1} dv$$

This integral can be found from tables. It is equal to $\pi/sin(\pi c)$.

Bibliography

This list of suggested books serves to expand one's knowledge in each of the areas under which each book is listed. The research articles quoted in the present book give a broader view of specific subjects treated in the text. They are not listed here but at the end of each chapter.

QUANTUM MECHANICS

1) A. P. French and Edwin F. Taylor, *An Introduction to Quantum Physics*, The MIT Introductory Physics Series, 1978.
   Although it seems introductory, it goes deeper than its title suggests. Written in the 1970's it is still in use today with a kindle edition.
2) Peter LHagelstein, Stephen D. Senturia, and Terry Orlando, *Introductory Applied Quantum and Statistical Mechanics*, Wiley, 2004.
   A recent book that includes chapters on statistical mechanics. Not so applied as it claims.
3) P. A.M. Dirac, *The Principles of Quantum Mechanics*, Oxford University Press, 1976.
   This is the standard book on the fundamental principles by one of the founders of Quantum Mechanics. Not easy because of the abstract notation, but probably the book with the clearest presentation of principles.

SOLID STATE PHYSICS AND CLASSICAL TRANSPORT THEORY

1) J.M. Ziman, *Principles of the Theory of Solids*, Cambridge University Press (2nd edition), 1972.
   A classic and authoritative textbook on the foundations of solid state theory but it does not contain much about Quantum Transport. Though written decades ago, it is still useful. It has been translated in many languages.
2) Giuseppe Grosso and Giuseppe Pastori Parraviani, *Solid State Physics*, Academic Press, 2000.
   A recent, comprehensive, and clear book.

DEVICES, MAINLY CLASSICAL APPROACH

1) YuanTaur and Tak H. Ning, *Fundamentals of Modern VLSI Devices* (2nd edition), Cambridge University Press, 2009.
   A comprehensive textbook mainly on devices and electronics.
2) Donald A. Neamen, *Semiconductor Physics and Devices* (4th edition), McGraw-Hill, 2011
3) Umesh K. Mishra and Jasprit Singh, *Device Physics and Design*, Springer, 2008.
   It gives a comprehensive but classical coverage of devices and offers a selection of a few Quantum Transport cases in the appendices.

QUANTUM TRANSPORT AND DEVICES

1) John H. Davies, *The Physics of Low-Dimensional Semiconductors*, Cambridge University Press, 1998.
   The book deals almost exclusively with Solid State Physics at the nano-level but contains a good account of Quantum Transport theory.

2) Vladimir V. Mitin, Viatcheslav A.Kochelap, and Michael A.Strossio, *Quantum Heterostructures*, Cambridge University Press, 1999.

A book with a good blend of Physics and Devices at the nano-level.

3) Mark Lundstrom, *Fundamental of Nanotransistors*, World Scientific Publishing, 2018.

A book by one of the main contributors in the field. It deals almost exclusively with the MOSFET but it will be a book to be consulted frequently.

Index

Note: Page numbers in *italics* and **bold** refer to figures and tables, respectively.